Multiplicative Complexity, Convolution, and the DFT

Michael T. Heideman

Multiplicative Complexity, Convolution, and the DFT

C.S. Burrus, Consulting Editor

Springer-Verlag
New York Berlin Heidelberg
London Paris Tokyo

Michael T. Heideman
Etak, Incorporated
Menlo Park, CA 94025
USA

Consulting Editor
Signal Processing and Digital Filtering

C.S. Burrus
Professor and Chairman
Department of Electrical and
    Computer Engineering
Rice University
Houston, TX 77251-1892
USA

Library of Congress Cataloging-in-Publication Data
Heideman, Michael T.
    Multiplicative complexity, convolution, and the DFT / Michael T.
Heideman ; consulting editor, C.S. Burrus.
        p.   cm.
    Originally presented as the author's thesis (Ph. D.—Rice
University) under title: Applications of multiplicative complexity
theory to convolution and the discrete Fourier transform.
    Bibliography: p.
    Includes index.
    1. Computational complexity.   2. Fourier transformations.
3. Convolutions (Mathematics)   I. Burrus, C. S.   II. Title.
QA267.H44   1988
511—dc19   88-16048

Camera-ready copy provided by the author.
Printed and bound by R.R. Donnelley and Sons, Harrisonburg, Virginia.
Printed in the United States of America.

9 8 7 6 5 4 3 2 1

ISBN 0-387-96810-5 Springer-Verlag New York Berlin Heidelberg
ISBN 3-540-96810-5 Springer-Verlag Berlin Heidelberg New York

Preface

This book is intended to be a comprehensive reference to multiplicative complexity theory as applied to digital signal processing computations. Although a few algorithms are included to illustrate the theory, I concentrated more on the development of the theory itself.

Howie Johnson's infectious enthusiasm for designing efficient DFT algorithms got me interested in this subject. I am grateful to Prof. Sid Burrus for encouraging and supporting me in this effort. I would also like to thank Henrik Sorensen and Doug Jones for many stimulating discussions.

I owe a great debt to Shmuel Winograd, who, almost singlehandedly, provided most of the key theoretical results that led to this present work. His monograph, *Arithmetic Complexity of Computations*, introduced me to the mechanism behind the proofs of theorems in multiplicative complexity, enabling me to return to his earlier papers and appreciate the elegance of his methods for deriving the theory. The second key work that influenced me was the paper by Louis Auslander and Winograd on multiplicative complexity of semilinear systems defined by polynomials. After reading this paper, it was clear to me that this theory could be applied to many important computational problems. These influences can be easily discerned in the present work.

I have attempted to include proofs of all the results in this book. This should make the book more self-contained, so that the reader can follow the proofs in the book itself, without needing to refer to papers from journals. On the other hand, it should be possible for a reader to skip over the proofs if the results are all that is desired.

This work was first printed as my PhD thesis at Rice University titled "Applications of multiplicative complexity theory to convolution and the discrete Fourier transform." This book includes all of the original thesis plus many more examples, a set of problems, several added sections containing new material, and corrections to any typographical errors I was able to find.

Lastly, I would like to thank my wife Edie for her support throughout the original composition and the final revision of this book.

Palo Alto, California   Michael T. Heideman
March 1988

Contents

CHAPTER 1

Introduction

1.1. An Overview of Multiplicative Complexity

From man's first encounters with the abstract concepts of numbers and arithmetic he has tried to simplify the numeric computations that extract desired information from numerical measurements. The abacus is a device invented long ago whose sole purpose is to simplify numerical computation. Beginning in the 16th Century, numerical analysis techniques were developed to simplify hand calculation. The advent of digital computers has inspired the development of new techniques for calculations that had been overlooked earlier because of the impossibility of manually performing them.

An effort began in the early 1950's to quantify the number of multiplication or division operations necessary to compute certain functions. The number of multiplications necessary to evaluate a polynomial at a fixed point was first considered. Later, in the 1960's, some results were obtained for products of polynomials, some specific matrix products, and the discrete Fourier transform. These results were mostly upper bounds on the number of required multiplications, obtained by inventing an algorithm, either heuristically or theoretically, which rarely enabled a lower bound to be derived for the same computations.

In the early 1970's, the theory of multiplicative complexity emerged as a new field of study, unifying some of the earlier results into a common framework. The number of multiplications/divisions necessary and sufficient to compute several particular systems was established. Rather than loosely bounding the complexity, an exact minimum number of multiplication/division operations is determined that will compute a system. For instance, it has been proven that 7 multiplications are necessary and sufficient to compute the product of two arbitrary 2×2 matrices.

In this book multiplicative complexity theory is applied to systems commonly encountered in digital signal processing. These include aperiodic and cyclic convolutions and the discrete Fourier transform. Winograd has derived some multiplicative complexity results for these systems when all inputs are indeterminate; in particular, exact expressions for the minimum number of required multiplications are known for one-dimensional aperiodic and cyclic convolution and the discrete Fourier transform (DFT) of a prime-length sequence. Lower bounds have also been derived for the multiplicative complexity of cyclic convolution when only one input sequence is

completely indeterminate. Proofs of these results are included herein for completeness, since the derivations have never before been collected in one place.

Recently, I and several others have shown how to evaluate the multiplicative
complexity of the DFT for sequence lengths that are powers of prime numbers
(including 2). These results have been extended to arbitrary sequence lengths in one
dimension. The multiplicative complexity of multidimensional cyclic convolution
has also been derived for any possible combination of lengths.

Another contribution of this book is an analysis of the multiplicative complexity
of convolution and the discrete Fourier transform when either the inputs have been
constrained or the outputs have been restricted. It is shown that only certain types of
input constraints reduce the multiplicative complexity of a system of aperiodic convolution. The discrete Fourier transform with symmetric or antisymmetric inputs is
investigated. The multiplicative complexity of computing certain subsets of the outputs of the aforementioned systems is analyzed, but general results for this type of
restriction are not as easily obtained as for the input constraints.

1.2. Why Count Only Multiplications and Divisions?

Why is it so important to count multiplications and divisions, while ignoring
other factors that contribute to the computational complexity of a system? Perhaps
the best answer to this question is simply that techniques have been developed for
evaluating multiplicative complexity, and hence this counting can be done. No
theory has yet been developed that accounts for all the factors influencing the computational complexity of a system. Most methods assume an algorithmic framework
through which some of these factors can be analyzed, but this is not particularly general since the framework itself has limited the means of computing the system, and it
is usually not possible to show that the most efficient algorithm lies within the framework.

The types of systems to which the theory of multiplicative complexity can be
applied are those in which a desired set of outputs is expressed exactly in terms of a
given set of inputs using only the field operations of addition, subtraction, multiplication, and division. Of these four operations, multiplication and division are intrinsi-
cally more difficult, generally requiring many additions/subtractions to be approxi-
mated to a desired precision in most fields of interest. Much of the computer
hardware built before the mid-1970's and most fixed-point processors today require
significantly more time to multiply or divide two numbers than to add or subtract
them.

Analysis of the multiplicative complexity of a system of equations provides a
definitive number as the minimum number of required multiplications/divisions,
which is useful information when evaluating known algorithms for computing the
system. The techniques used to evaluate the multiplicative complexity of a system
frequently   yield   algorithms   that   realize   the   minimum   number   of

multiplications/divisions, or provide insights into the development of suboptimal algorithms that may be more efficient than the known algorithms.

The number of multiplications or divisions used to compute a given system can often be reduced by applying the distributive, commutative, and associative laws in the given field, particularly by applying these laws to exploit inherent algebraic properties of the system. For large systems, the algorithms that realize the minimum number of multiplications/divisions often require excessive numbers of additions, and can be extremely susceptible to numerical errors when implemented on standard computing hardware. It is often possible to decompose a large system into smaller systems, each lacking the problem of excessive additions when individually implemented with the minimum number of multiplications/divisions, to obtain suboptimal algorithms that are practical and less likely to exhibit unacceptable computational errors.

This book concentrates exclusively on the multiplicative complexity of algorithms. Some research into the additive complexity of systems was undertaken in the early 1970's by Morgenstern. More recently, Pan investigated additive, logical, and bit complexities of systems. These results tend to be very loose bounds and sometimes have strange caveats, such as "this lower bound applies only when all multiplications involve constants with magnitude less than or equal to one."

The primary reason that we are interested in the number of multiplications and divisions is that we want to be able to compute the outputs of a system in as little time as possible. If the number of additions or nontrivial rational multiplications increases substantially when multiplications are minimized, then this original problem has not been solved. A unified theory is needed that will specify an algorithm that computes a given system on a specific piece of computing hardware to a specified precision in the least amount of time. As this unified theory is not yet (and may never be) reality, we must be content to solve small parts of the overall puzzle, slowly working toward the complete solution.

1.3. Organization

The remaining chapters describe and analyze the various types of systems encountered, using an abstract algebraic approach originally due to Winograd. Placing these systems into this algebraic framework requires notation that is far more familiar to mathematicians than engineers. An attempt has been made to shed some of the abstraction by including frequent examples of applications of the abstract principles.

Chapter 2 presents the basis for the theory of multiplicative complexity and shows how to apply this theory to bilinear systems. *Semilinear systems* are defined and the multiplicative complexity of several specific semilinear systems is analyzed. The concept of a *substitution* is introduced. The most important results of this chapter are the *row-rank theorem* and the *column-rank theorem*.

Chapter 3 applies the theory of multiplicative complexity to several types of systems of polynomial multiplication, including systems equivalent to aperiodic and cyclic convolution. Some results on polynomial multiplication modulo irreducible polynomials are applicable to the analysis of the multiplicative complexity of the discrete Fourier transform, as are the extensions to general polynomials, polynomials in several variables, and multiplication by several polynomials. A large part of this chapter is devoted to the presentation of some general results of Auslander and Winograd on semilinear systems defined by polynomials. The multiplicative complexity of multidimensional cyclic convolution in an arbitrary number of dimensions is derived.

Chapter 4 analyzes polynomial products with constraints. An important theorem is proven that specifies exact conditions an input constraint must satisfy if it is to reduce the multiplicative complexity of a system. This theorem is then used to analyze the multiplicative complexity of polynomial products with various symmetries. A theory is also proposed for some types of restrictions on the outputs of systems of polynomial multiplication, but the results are less general because of the different structure of this problem.

Chapter 5 applies the results of Chapter 3 to determine the multiplicative complexity of the one-dimensional discrete Fourier transform for all possible lengths. Examples are included to demonstrate the application of the theory.

Chapter 6 extends the analysis of the multiplicative complexity of the discrete Fourier transform to include constraints on the input and computation of subsets of the output. These special situations include both input and output pruning and transforms of symmetric and antisymmetric sequences. The multiplicative complexity of the discrete Hartley transform and discrete cosine transform are derived as special cases.

CHAPTER 2

Multiplicative Complexity of Linear and Bilinear Systems

Multiplicative complexity theory is a field that has existed for only about thirty years. This chapter begins by highlighting some of the major work in this field that has influenced recent research into the multiplicative complexity of convolution and the discrete Fourier transform. A mathematical framework is then established for the study of arithmetic algorithms and their multiplicative complexity, emphasizing applications to bilinear systems. Semilinear systems are introduced to facilitate the study of bilinear systems with one fixed set of inputs, such as a digital filter or the discrete Fourier transform. The important concepts of direct products and direct sums of systems are also introduced.

2.1. Historical Perspective

The rigorous development of the theory of multiplicative complexity of arithmetic algorithms has a short history, beginning with the development of some heuristic algorithms for polynomial evaluation in the mid-1950's [25, 29, 30, 31]. Schemes for multiplication of large integers [14, 22, 39], that were later extended to the multiplication of polynomials, soon followed. In 1969, Strassen [36] proposed an algorithm for multiplication of 2×2 matrices that, when iterated, produced new algorithms for multiplication of matrices of arbitrary size that use fewer multiplications than the straightforward method. These methods and the fast Fourier transform (FFT) of Cooley and Tukey [15] provided new upper bounds on the number of multiplications sufficient to compute certain systems. These successes spurred the development of methods, beginning in the early 1970's, for determining realizable lower bounds on the number of multiplications necessary to compute certain systems [10, 17, 32, 40, 41, 43, 45]. In this chapter, some of these methods are presented as theorems that will then be used extensively in the following chapters.

A notation must first be established to simplify the discussion of both multiplicative complexity and bilinear systems. Much of this notation is borrowed from Winograd [47] since he has been a main contributor to the theory. The following sections of this chapter present the fundamental concepts that are needed to determine the multiplicative complexity of convolution and the discrete Fourier transform.

2.2. Definitions and Basic Results

An arithmetic algorithm is a sequence of steps that exactly computes a desired set of outputs from the given set of inputs. This definition precludes algorithms that approximately compute an output or set of outputs, such as procedures that minimize the value of a function, find roots, etc.

At this point, it may be helpful to refresh your mind on certain algebraic concepts and notation. If you do not feel comfortable with sets, groups, fields, rings, extension fields, etc., I recommend reviewing a basic text on algebra before proceeding, although most of these constructions should be fairly clear from their context and from the examples.

Some multiplications and divisions are trivial, such as multiplication or division by unity. To prevent these trivial multiplications and divisions from being counted, a *ground set G*, sometimes called the *field of constants*, is specified. Any multiplication or division by an element of G is not included in the count of multiplication/division operations used in an algorithm. G is often chosen to be Q, the field of rational numbers, since multiplication by an integer can be carried out with a finite number of additions and/or subtractions, and division by an integer can be avoided by scaling the outputs so that each has the same integer denominator.

The *base set B* is the set of inputs to an algorithm, including all elements of G as a subset, plus a set of indeterminate values that are not in G. Later we will find it convenient to define a field $F = G(x_1, x_2, \ldots, x_m)$, and then let $B = F \cup \{y_1, y_2, \ldots, y_n\}$ where $\{x_1, x_2, \ldots, x_m\}$ and $\{y_1, y_2, \ldots, y_n\}$ are sets of distinct indeterminates. This is a useful construction for bilinear systems when one of the sets of indeterminates (the x_i's) is known á priori, allowing any operations that involve only elements of this set to be done in advance, so that these precomputed quantities can be considered as inputs to the algorithm.

The field $H = F(y_1, y_2, \ldots, y_n)$ contains all possible results of (repeated) application of the field operations to the base set B. The values to be computed are the set $Z = \{z_1, z_2, \ldots, z_t\} \subseteq H$, and the computation of these output values from a given B will be called the *system Z*.

Definition 2.1. *Let G be a field, and let $F \supseteq G$ be a field. Let $H = F(y_1, y_2, \ldots, y_m)$ be a purely transcendental extension field of F. An algorithm A (over the set B) is a finite sequence $h_1, h_2, \ldots, h_n$ of elements of H such that either $h_i \in B$ or there exist $j, k < i$ such that $h_i = h_j \# h_k$, where $\#$ is a field operation.*

The algorithm A imposes an ordering on the elements of H such that each successive element must either be an element of the base set or must have been computed from two previous elements using the field operations of addition, subtraction, multiplication, or division.

Definition 2.2. *An algorithm $A = (h_1, h_2, \ldots, h_n)$ over B is said to compute $Z = \{z_1, z_2, \ldots, z_t\}$ if for every i $(1 \leq i \leq t)$ there exists j such that $z_i = h_j$.*

We have now defined an algorithm and shown when an algorithm computes the outputs of a system. We must define what is meant by a *multiplication/division (m/d) step* in order to evaluate the multiplicative complexity of a given system. Of the four field operations, multiplication, division, addition, and subtraction, the last two are never considered to be m/d steps and the first two are considered m/d steps unless at least one operand is an element of the field of constants, G. The following definition formalizes this concept.

Definition 2.3. *Let $A = (h_1, h_2, \ldots, h_n)$ be an algorithm over B. An h_i is a non-m/d step if it is either equal to an input, equal to the sum (or difference) of two steps preceding it, or equal to the product of a previous step and an element of G. Any h_i that does not satisfy one of these conditions is an m/d step.*

The number of m/d steps in an algorithm A is denoted by $\mu(A)$, or $\mu_B(A)$ or $\mu_B(A; G)$, if the dependence on B or G is to be emphasized.

Definition 2.4. *The multiplicative complexity of the system $Z = \{z_1, z_2, \ldots, z_t\}$ denoted by $\mu(Z)$ or $\mu(z_1, z_2, \ldots, z_t)$ is*

$$\mu(Z) = \min_A \mu(A)$$

where A ranges over all algorithms over B (with G as the field of constants) computing Z.

$\mu(Z)$ is obviously the minimum number of m/d steps, as specified in Definition 2.3, necessary to compute Z. The multiplicative complexity is sometimes denoted by $\mu_B(Z; G)$ to emphasize the dependence on B and G. The notations for the number of m/d steps in an algorithm and the multiplicative complexity of a system are identical but distinguishable by identifying the first argument in the parentheses as an algorithm or set of outputs.

Definition 2.5. *The linear span (over G) of B is the set of all elements expressible as $\sum g_i b_i$, where $g_i \in G$ and $b_i \in B$, and is denoted by $L_G(B)$. When $B = F \cup \{y_1, y_2, \ldots, y_n\}$, then each element of $L_G(B)$ can be expressed as $f + \sum_{i=1}^{n} g_i y_i$ where $f \in F$ and $g_i \in G$, $i = 1, 2, \ldots, n$.*

The following example will clarify the preceding definitions.

Example 2.1. Let the ground set be $G = Q$, the field of rational numbers, and let $Z = \{z_0, z_1\}$ be the system defined by $z_0 = \pi y_0$, $z_1 = e y_1$, where y_0 and y_1 are indeterminate elements of some larger field, π is the ratio of the circumference to the diameter of a circle, and e is the base used in computing natural logarithms. Since the values of π and e are known, it may be possible to precompute certain functions involving them, thus it will be assumed that $F = Q(\pi, e)$, a transcendental extension field of the rationals. The quantities that may be precomputed are all of the form $\sum_{i=-\infty}^{\infty} \sum_{j=-\infty}^{\infty} g_{ij} \pi^i e^j$, where $g_{ij} \in Q$ for $-\infty < i, j < \infty$. Examples of elements of F include $e\pi$, π/e, $1+\pi+e$, $3\pi^2$, $-7/2$, etc. The convergence of the infinite sum is not an issue, since for all practical cases only a finite number of the g_{ij} are nonzero.

The base set is $B = F \cup \{y_0, y_1\}$, and thus $L_G(B)$ is composed of all numbers expressible as $f + g_0 y_0 + g_1 y_1$, where $f \in Q(\pi, e)$ and $g_0, g_1 \in Q$. The set of all attainable results is $H = F(y_0, y_1) = Q(\pi, e, y_0, y_1)$, consisting of all elements of the form $\sum_{i=-\infty}^{\infty} \sum_{j=-\infty}^{\infty} \sum_{k=-\infty}^{\infty} \sum_{l=-\infty}^{\infty} g_{ijkl} \pi^i e^j y_0^k y_1^l$, where $g_{ijkl} \in Q$, $-\infty < i, j, k, l < \infty$.

One possible algorithm for computing this system is $A = (h_1, h_2, \ldots, h_6)$, where $h_1 = \pi$, $h_2 = e$, $h_3 = y_0$, $h_4 = y_1$, $h_5 = h_1 \cdot h_3$, $h_6 = h_2 \cdot h_4$. This algorithm does indeed compute the system since $z_0 = h_5$ and $z_1 = h_6$. Only two m/d steps, h_5 and h_6, were needed, thus $\mu(A) = 2$. It can be shown using one of several theorems proved in the next section that $\mu(Z) = 2$.

Another algorithm that computes Z is $\bar{A} = (h_1, h_2, \ldots, h_8)$, where $h_1 = \pi$, $h_2 = \pi + e$, $h_3 = y_0$, $h_4 = y_1$, $h_5 = h_1 \cdot h_3$, $h_6 = h_2 \cdot h_4$, $h_7 = h_1 \cdot h_4$, $h_8 = h_6 - h_7$. In $\bar{A}$, $z_0 = h_5$, $z_1 = h_8$, and $\mu(\bar{A}) = 3$ since h_5, h_6, and h_7 are m/d steps. This algorithm demonstrates that any elements of B can be assigned in the algorithm since the step h_2, although it appears to be an addition step, is really an assignment to an element of F (and B).

From Definition 2.4 it is obvious that if each of the outputs is equal to an input then the multiplicative complexity is zero. There are other systems with a multiplicative complexity of zero, and it is possible to completely specify all systems that require no m/d steps for a given B and G. Clearly, adding an element of $L_G(B)$ as a step in an algorithm A does not affect the multiplicative complexity of A. Adding or subtracting two elements of $L_G(B)$ yields another element of $L_G(B)$. Similarly, multiplying an element of $L_G(B)$ by an element of G also yields an element of $L_G(B)$. Therefore, by induction, if each step is a non-m/d step, then each step, and in particular each output, must be in $L_G(B)$. Conversely, if any output requires a m/d step, it is not in $L_G(B)$, leading to the following lemma.

Lemma 2.1. *The multiplicative complexity of a system $Z = \{z_1, z_2, \ldots, z_t\}$ is equal to zero if and only if $z_i \in L_G(B)$, $i = 1, 2, \ldots, t$.*

Let H' be the quotient space $H/L_G(B)$. The natural vector space homomorphism will be denoted by $r : H \to H'$. The homomorphism r is sometimes referred to as reduction modulo $L_G(B)$. It will be convenient to view the outputs and the steps in A as (representatives of) elements of H'.

Theorem 2.1. [47] *Let* $A = (h_1, h_2, \ldots, h_n)$ *be an algorithm over* B, *and let* $h(1), h(2), \ldots, h(s)$ *be the m/d steps of* A; *then for each* $i = 1, 2, \ldots, n$, $r(h_i) \in L_G(r(h(1)), r(h(2)), \ldots, r(h(s)))$.

Proof. Suppose A has only one m/d step $h(1)$. All other steps in A may be expressed as either $h_i = h_j \pm h_k$, $j, k < i$, or $h_i = gh_j$, $j < i$, $g \in G$, or $h_i \in B$. Therefore all steps preceding $h(1)$ must be of the form $h_i = \sum g_{ij} b_j$, $g_{ij} \in G$, $b_j \in B$, and all steps after $h(1)$ are of the form $h_i = \sum g_{ij} b_j + g'h(1)$, $g_{ij}, g' \in G$, $b_j \in B$, and thus $r(h_i) \in L_G(r(h(1)))$, $i = 1, 2, \ldots, n$.

Suppose that $r(h_i) \in L_G(r(h(1)), r(h(2)), \ldots, r(h(s-1)))$, $i = 1, 2, \ldots, n-1$, and an m/d step $h_n = h(s)$ is appended to A. Then $r(h_i) \in L_G(r(h(1)), r(h(2)), \ldots, r(h(s-1))) \subseteq L_G(r(h(1)), r(h(2)), \ldots, r(h(s)))$, $i = 1, 2, \ldots, n-1$, and clearly $r(h_n) = r(h(s)) \in L_G(r(h(1)), r(h(2)), \ldots, r(h(s)))$, therefore $r(h_i) \in L_G(r(h(1)), r(h(2)), \ldots, r(h(s)))$, $i = 1, 2, \ldots, n$.

Similarly, if a non-m/d step h_n is appended to an algorithm A having s m/d steps, then it is either $h_n = h_j \pm h_k$, $j, k < n$, or $h_n = gh_j$, $g \in G$, $j < n$, or $h_n \in B$. If $h_n = h_j \pm h_k$, then $r(h_n) = r(h_j) \pm r(h_k) \in L_G(r(h(1)), r(h(2)), \ldots, r(h(s)))$. If $h_n = gh_j$, then $r(h_n) = gr(h_j) \in L_G(r(h(1)), r(h(2)), \ldots, r(h(s)))$. Lastly, if $h_n \in B$, then $r(h_n) = 0$. By induction on s, $r(h_i) \in L_G(r(h(1)), r(h(2)), \ldots, r(h(s)))$, $i = 1, 2, \ldots, s$. ∎

The following key theorem can now be proven, allowing lower bounds to be established for the multiplicative complexity of many systems. This theorem states that the number of m/d steps necessary to compute a set of outputs from a given set of inputs is greater than or equal to the dimension of the linear span (over G) of the outputs.

Theorem 2.2. [47] $\mu_B(Z) \geq \dim L_G(r(z_1), r(z_2), \ldots, r(z_t))$ *where* $Z = \{z_1, z_2, \ldots, z_t\}$.

Proof. Let $A = (h_1, h_2, \ldots, h_n)$ be an algorithm computing Z such that the m/d steps of A are $h(1), h(2), \ldots, h(s)$. Theorem 2.1 states that $r(h_i) \in L_G(r(h(1)), r(h(2)), \ldots, r(h(s)))$, $i = 1, 2, \ldots, n$, and thus $L_G(r(z_1), r(z_2), \ldots, r(z_t)) \subseteq L_G(r(h(1)), r(h(2)), \ldots, r(h(s)))$. Therefore $\dim L_G(r(z_1), r(z_2), \ldots, r(z_t)) \leq \dim L_G(r(h(1)), r(h(2)), \ldots, r(h(s))) \leq s$. Had A been an algorithm satisfying $\mu_B(A) = \mu_B(Z)$, then $s = \mu_B(Z)$, and thus $\mu_B(Z) \geq \dim L_G(r(z_1), r(z_2), \ldots, r(z_t))$. ∎

Corollary 2.1. *If* t *output values are being computed that are linearly independent (over* G) *in* H', *then at least* t *multiplications are required to compute the set of outputs.*

An algorithm that computes the outputs of a system using the minimum number of m/d steps is called a *minimal algorithm*. Occasionally some of the outputs can be directly computed from the inputs using a single m/d step independent of the other outputs. The next theorem shows that a minimal algorithm exists that computes each element of this subset directly.

Theorem 2.3. [47] *Let* $Z = \{z_1, z_2, \ldots, z_l\} \subseteq H$ *and let* $Z' = \{z_1, z_2, \ldots, z_k\} \subseteq Z$ *be such that* $r(z_1), r(z_2), \ldots, r(z_k)$ *are linearly independent over* G *and each* $z_i \in Z'$ *satisfies* $\mu_B(z_i; G) = 1$. *There exists an algorithm* A *over* B *computing* Z *such that* $\mu_B(A; G) = \mu_B(Z; G)$ *and the first* k *m/d steps of* A *are the elements of the set* Z'.

Proof. Let $A' = (h_1, h_2, \ldots, h_n)$ be a minimal algorithm for computing $Z = \{z_1, z_2, \ldots, z_l\}$ and let $h(1), h(2), \ldots, h(s)$ be the m/d steps of A'. Assume, without loss of generality, that $Z' = \{z_1, z_2, \ldots, z_k\}$. By Theorem 2.1, $r(z_k) \in L_G(r(h(1)), r(h(2)), \ldots, r(h(s)))$. Let q be the smallest integer such that $r(z_k) \in L_G(r(h(1)), r(h(2)), \ldots, r(h(q)))$. This output value can be expressed as $z_k = \sum_{i=1}^{q} g_i h(i) + \sum g_i' b_i$ where $g_i, g_i' \in G$, $b_i \in B$, and by the minimality of q, $g_q \neq 0$. The algorithm A' can be modified by inserting an initial sequence of steps that compute z_k using only one m/d step, and then replacing the m/d step $h(q)$ by a sequence of non-m/d steps using the relation $h(q) = g_q^{-1}(z_k - \sum_{i=1}^{q-1} g_i h(i) - \sum g_i' b_i)$. The new algorithm has the same number of m/d steps as A' and contains all steps in A', and hence must also compute Z. If this process is repeated for $z_{k-1}, z_{k-2}, \ldots, z_1$, then A' will have been converted into the algorithm A satisfying the conditions of the theorem. ∎

Example 2.2. Consider the system $Z = \{\pi y_0, e y_1\}$ introduced in Example 2.1. Recall the algorithms

$$A = (h_1 = \pi, h_2 = e, h_3 = y_0, h_4 = y_1, h_5 = h_1 \cdot h_3, h_6 = h_2 \cdot h_4)$$

and

$$\tilde{A} = (h_1 = \pi, h_2 = \pi + e, h_3 = y_0, h_4 = y_1,$$
$$h_5 = h_1 \cdot h_3, h_6 = h_2 \cdot h_4, h_7 = h_1 \cdot h_4, h_8 = h_6 - h_7).$$

In both A and $\tilde{A}$ the first four steps are in $L_G(B)$ and thus for both algorithms $r(h_i) = 0$, $i = 1, 2, 3, 4$. The remaining steps have no linear component over B for either algorithm and consequently for these steps $r(h_i) = h_i$.

In the algorithm A, the space $L_G(r(h_5), r(h_6))$ is composed of all elements expressible as $g_0 \pi y_0 + g_1 e y_1$ for $g_0, g_1 \in G$. Clearly the mappings $r(h_i)$ of the first four steps are contained in this space by letting $g_0 = g_1 = 0$, and since $r(h_5)$ and $r(h_6)$ are basis vectors of the space, they are naturally included also, validating Theorem 2.1

for this algorithm.

The algorithm $\tilde{A}$ has three m/d steps, and the relevant space is $L_G(r(h_5), r(h_6), r(h_7))$. This space has elements of the form $g_0\pi y_0 + g_1(\pi+e)y_1 + g_2\pi y_1$, and since the basis vectors are independent over G the dimension of the space is three. The validity of Theorem 2.1 can be demonstrated for this algorithm, as for A, since the first four steps are the zero vector, the next three steps are the basis vectors, and the final step is the vector $[0 \; 1 \; -1]^T$, where the natural basis is used.

Theorem 2.2 states that $\mu_B(Z) \geq \dim L_G(\pi y_0, ey_1)$, and for indeterminates y_0 and y_1, no $g_0, g_1 \in G$ exist such that $g_0\pi y_0 + g_1 ey_1 = 0$, thus $\dim L_G(\pi y_0, ey_1) = 2$ and $\mu_B(Z) \geq 2$. Corollary 2.1 also applies to this example since 2 independent outputs are being computed. Both outputs require only one m/d step each and are independent, thus Theorem 2.3 applies, and A is an example of an algorithm that does indeed compute both outputs directly.

2.3. Semilinear Systems

The preceding development has given a general framework for the evaluation of the multiplicative complexity of systems. This analysis will now be specialized to the class of semilinear systems. A *semilinear system* is defined to be the computation of $Z = \{z_1, z_2, \ldots, z_t\}$ where $z_i = \sum_{j=1}^{n} \phi_{ij} y_j$, $i = 1, 2, \ldots, t$. The set of inputs is $B = F \cup Y$, where $Y = \{y_1, y_2, \ldots, y_n\}$ are indeterminate elements of H, and F is the extension field of G by the ϕ_{ij}'s. In matrix form this system is $z = \Phi y$, where z is the vector $z = [z_1 \; z_2 \; \cdots \; z_t]^T$, y is the vector $y = [y_1 \; y_2 \; \cdots \; y_n]^T$, and Φ is the $t \times n$ matrix with ϕ_{ij} the entry in the ith row and j^{th} column.

The space over G whose elements are n-tuples of elements of F will be called V, and the corresponding space of t-tuples of elements of F will be denoted by W. Each row of Φ is thus an element of V and each column of Φ is an element of W. Let V' be the quotient space $V' = V/G^n$, in which each row of Φ represents an element. Define $\rho_r(\Phi)$ to be the dimension of the linear space over G spanned by the rows of Φ as elements of V', and define $\rho_c(\Phi)$ to be the dimension of the linear space over G spanned by the columns of Φ as elements of the quotient space $W' = W/G^t$. The following theorem is simply a rewording of Theorem 2.2 using this notation.

Theorem 2.4. [17,47] *Let* $Z = \Phi y$ *be a semilinear system and* $\rho_r(\Phi)$ *the row rank of* Φ *over* G, *then* $\mu_B(\Phi y) \geq \rho_r(\Phi)$.

Proof. Let Z in Theorem 2.2 be the set composed of the entries in the vector $z = \Phi y$. The entries of y are independent over G, thus each of the $\rho_r(\Phi)$ independent rows of Φ (in V') will generate an independent output (over $L_G(B)$). Therefore $\dim L_G(r(z_1), r(z_2), \ldots, r(z_t)) \geq \rho_r(\Phi)$, which by the transitivity of inequality and

Theorem 2.2 yields $\mu_B(\Phi y) \geq \rho_r(\Phi)$. ∎

A similar result can be proven for the column rank of Φ, but before presenting this theorem, it will be useful to define the concept of a *substitution*. Intuitively, a substitution replaces one of $\{y_1, y_2, \ldots, y_n\}$ with a linear combination of the other inputs to the system. The resulting system may require fewer m/d steps than the original system. The partial mapping α^* is included in the definition to eliminate the possibility of division by zero when a substitution is made in an algorithm.

Definition 2.6. *A substitution is a mapping* $\alpha:\{y_1, y_2, \ldots, y_n\} \to L_G(B)$. *This mapping can be extended uniquely to a homomorphism* $\bar{\alpha}: F[y_1, y_2, \ldots, y_n] \to F[y_1, y_2, \ldots, y_n]$ *such that every* $f \in F$ *remains fixed. A partial mapping* α^* *can be defined as* $\alpha^*: F[y_1, y_2, \ldots, y_n] \to F[y_1, y_2, \ldots, y_n]$ *such that* $\alpha^*(a/b) = \bar{\alpha}(a)/\bar{\alpha}(b)$ *when* $\alpha(b) \neq 0$ *for* $a, b \in F(y_1, y_2, \ldots, y_n)$. *A substitution is called a specialization of* y_j *when* $\alpha(y_i) = y_i$ *for* $i \neq j$ *and* $\alpha(y_j) = f + \sum_{k \neq j} g_k y_k$ *for* $f \in F$ *and* $g_k \in G$.

Definition 2.7. *A substitution* α *is compatible with an algorithm* A *if every step of* A *is expressible as* a/b, $\bar{\alpha}(b) \neq 0$, *and* $a, b \in F[y_1, y_2, \ldots, y_n]$.

Theorem 2.5. [40] *Let* $Z = \Phi y$ *be a semilinear system and* $\rho_c(\Phi)$ *the column rank of* Φ *over* G, *where* $|G| = \infty$, *then* $\mu_B(\Phi y) \geq \rho_c(\Phi)$.

Proof. Let Z of Theorem 2.2 be as in Theorem 2.4. Assume initially that $\rho_c(\Phi) = 1$, so that at least one column of Φ exists that is not in G^t. Assume without loss of generality that the jth column of Φ is not in G^t. At least one entry in the jth column, assumed to be ϕ_{ij}, is not in G. Therefore the ith row of Φ is not in G^n, and by the row rank theorem, Theorem 2.4, we conclude that $\mu_B(\Phi y) \geq 1$, proving the assertion for $\rho_c(\Phi) = 1$.

Suppose that the assertion holds for $\rho_c(\Phi) = u$, and let A be a minimal algorithm computing Φy, where Φ has $u+1$ columns that are G-linearly independent in W'. Let $h(1)$ be the first m/d step of A. All steps preceding $h(1)$ must be of the form $h = \sum_{i=1}^{n} g_i y_i + f$, $g_i \in G$, $i = 1, 2, \ldots, n$, $f \in F$. Either $h(1) = h \cdot h'$ or $h(1) = h \div h'$ for two steps h and h' in A preceding $h(1)$. Let $h = \sum_{i=1}^{n} g_i y_i + f$ and $h' = \sum_{i=1}^{n} g_i' y_i + f'$, where $g_i, g_i' \in G$, $i = 1, 2, \ldots, n$, and $f, f' \in F$. At least one g_i or g_i' is not zero, since if they were all zero then $h(1) \in F \subseteq L_G(B)$, contradicting the assumption that $h(1)$ is an m/d step. Assume, with no loss of generality, that $g_n \neq 0$.

Choose $g \in G$ such that if $g_n^{-1}(g - f - \sum_{i=1}^{n-1} g_i y_i)$ is substituted for y_n, then no steps in A will require division by zero. Such a g can always be selected since each step in A

can be expressed as a ratio of polynomials in y_n, which must have a finite number of roots because the number of steps in A is finite. Therefore only a finite number of possible substitutions for y_n will cause a denominator polynomial to be zero, and since G has an infinite number of elements, g can always be chosen such that all denominators are nonzero. After making the above substitution, an algorithm A' results that computes the system $\Phi'y'+v$ where $\phi'_{ij} = \phi_{ij} - g_n^{-1} g_j \phi_{in}$ $(1 \le i \le t,$ $1 \le j \le n-1)$, $y' = [y_1 \, y_2 \, \cdots \, y_{n-1}]^T$, and $v = (g-f)g_n^{-1}[\phi_{1n} \phi_{2n} \, \cdots \, \phi_{tn}]^T$. The m/d step $h(1)$ in A has been converted to a non-m/d step in A', and since the rest of the algorithm is identical, then the system Φy must require one more m/d step than $\Phi'y'+v$. The number of linearly independent columns of Φ' in W' must be at least u, consequently by the induction hypothesis at least u m/d steps are necessary to compute $\Phi'y'+v$. Therefore at least $u+1$ m/d steps are necessary to compute Φy and by induction at least $\rho_c(\Phi)$ m/d steps are necessary to compute Φy in general. ■

Theorems 2.4 and 2.5 are extremely valuable in the study of systems such as the discrete Fourier transform (DFT) that are linear in a set of indeterminates, and in which the matrix multiplying this vector of indeterminates consists of elements of H that are known exactly for all applications of this system. This contrasts with the general case where the elements of the matrix are also indeterminates, such as general matrix vector multiplication.

In the proof of Theorem 2.5, a substitution was performed that eliminated one m/d step from an algorithm. Some care had to be taken to insure that this substitution did not cause a division by zero anywhere in the algorithm. The following theorem guarantees the existence of a specialization that is compatible with a given system and such that the new system requires fewer m/d steps than the original system.

Theorem 2.6. [45] *Let G be an infinite field and $\{z_1, z_2, \ldots, z_l\}$ a semilinear system such that $\mu_B(z_1, z_2, \ldots, z_l) \ge 1$. A compatible specialization α of y_j exists such that*

$$\mu_B(\alpha^*(z_1), \alpha^*(z_2), \ldots, \alpha^*(z_l)) \le \mu_B(z_1, z_2, \ldots, z_l) - 1.$$

Proof. Use the substitution from Theorem 2.5, $g_n \alpha(y_n) = g - f - \sum\limits_{i=1}^{n-1} g_i y_i$. This substitution is compatible with the algorithm A since G is infinite and there are only a finite number of roots to each denominator polynomial in any step of A. The proof of Theorem 2.5 shows that this specialization also reduces the multiplicative complexity of the system by at least one m/d step. ■

The specialization of Theorem 2.6 can be repeated for several of the y_i, yielding the following corollary.

Corollary 2.2. [45] *For every algorithm A over B computing $\{z_1, z_2, \ldots, z_l\}$ there exists an algorithm A' over $B' = B \cup \{z_1, z_2, \ldots, z_l\}$ such that $\mu(A') \le \mu(A) - d(l)$, where $d(l) = \dim L_G(r(z_1), r(z_2), \ldots, r(z_l))$.*

Theorem 2.7. [45] *Let G be an infinite field, and* $\{z_1, z_2, \ldots, z_t\}$ *a semilinear system with* $z_i = f_i y_j$, $i = 1, 2, \ldots, l$. *There exists a compatible specialization* α *of* y_j *such that*

$$\mu_B(\alpha^*(z_{l+1}), \alpha^*(z_{l+2}), \ldots, \alpha^*(z_t)) \le \mu_B(z_1, z_2, \ldots, z_t) - d(l)$$

where

$$d(l) = \dim L_G(r(z_1), r(z_2), \ldots, r(z_l))$$
$$= \dim L_G(r(f_1), r(f_2), \ldots, r(f_l)).$$

Proof. Let A be a minimal algorithm computing $\{z_1, z_2, \ldots, z_t\}$. Assume without loss of generality that $j = 1$ and thus $y_j = y_1$. Every step of A can be expressed as $h_i = a_i/b_i$ where a_i and b_i are multivariate polynomials in $y_1, y_2, \ldots, y_n$. Since A has a finite number of steps and each denominator polynomial b_i has finitely many roots when viewed as a polynomial in y_1, there are only a finite set of elements of $F[y_2, y_3, \ldots, y_n]$ (the ring of polynomials with coefficients in F and indeterminates $y_2, y_3, \ldots, y_n$) that are roots of some denominator polynomial in A. Since G has infinitely many elements, choose g that is not a root of any polynomial $b_i(y_1)$ in A and let α be the specialization of y_1 defined by $\alpha(y_1) = g$ and $\alpha(y_i) = y_i$ for $i \ne 1$. Since no denominators become zero, α is compatible with $\{z_1, z_2, \ldots, z_t\}$ and clearly $\alpha^*(z_i) = gf_i \in L_G(B)$, $i = 1, 2, \ldots, l$. Corollary 2.2 states that for every algorithm A over B an algorithm A' over $B \cup \{z_1, z_2, \ldots, z_l\}$ exists for which $\mu(A') \le \mu(A) - d(l)$. If A had been a minimal algorithm for computing $\{z_1, z_2, \ldots, z_t\}$, then Corollary 2.2 implies that $\mu_B(z_{l+1}, z_{l+2}, \ldots, z_t) \le \mu_B(z_1, z_2, \ldots, z_t) - d(l)$. ∎

2.4. Quadratic and Bilinear Systems

Semilinear systems are a subclass of bilinear systems which are in turn a subclass of quadratic systems. Some general properties of bilinear and quadratic systems are presented in this section. The first important property is that a minimal algorithm using no division steps always exists for these systems. Commutative and noncommutative algorithms are then discussed. Lastly, direct products and direct sums are introduced as a means of creating large systems from smaller systems, or more importantly, for decomposing large systems into smaller systems.

2.4.1. Properties of Quadratic Systems

A *quadratic system* is a system of the form

$$z_k = \sum_{i=1}^{r} \sum_{j=1}^{i} g_{ijk} x_i x_j, \quad k = 1, 2, \ldots, t,$$

where the g_{ijk}'s are elements of G and the x_i's are indeterminate elements of H. A *bilinear system* is a special case of a quadratic system in which the indeterminates have been partitioned into the two sets $\{x_i; i = 1, 2, \ldots, r\}$ and $\{y_j; j = 1, 2, \ldots, s\}$, and each of the outputs z_k, $k = 1, 2, \ldots, t$, is a bilinear function of the elements of these two input sets. A general system of bilinear forms is then

$$z_k = \sum_{i=1}^{r} \sum_{j=1}^{s} g_{ijk} x_i y_j, \quad k = 1, 2, \ldots, t$$

where the g_{ijk}'s are elements of G and the x_i's and y_j's are indeterminate elements of H.

The following theorem shows two important features of minimal algorithms for bilinear systems. The first is that for any bilinear system there exists a minimal algorithm in which every m/d step is independent of any previous m/d steps. The second feature is that there exists a minimal division-free algorithm for computing any bilinear system. This suggests that in the study of the multiplicative complexity of bilinear systems we need only consider division-free algorithms in which no m/d step depends on a previous m/d step. This does not, in general, exclude the possibility that minimal algorithms exist in which one or more of the m/d steps involves division ([16] contains a counterexample), or in which an m/d step depends on one or more previous m/d steps. A more general theorem due to Strassen [37] concerning the conversion of essential division steps into multiplication steps, states that a system of degree d, requiring t m/d steps may be computed using at most $t_d = d(d-1)t/2$ essential multiplication steps. We prove this for the case $t = 2$.

Theorem 2.8. [47] *Let S be a system of quadratic forms. If G has infinitely many elements then a minimal algorithm A computing S exists such that each m/d step of A is of the form $h(i) = M_i(x)N_i(x)$ where $M_i(x)$ and $N_i(x)$ are both linear in x.*

Proof. Let S be the system $z_k = \sum_{i=1}^{r} \sum_{j \leq i} g_{ijk} x_i x_j$, $k = 1, 2, \ldots, t$. Let $A' = (h'_1, h'_2, \ldots, h'_n)$ be a minimal algorithm computing S. Each step of A' is a rational multivariate polynomial with coefficients in G. The algorithm A' will be modified to the algorithm A satisfying the conditions of the theorem by representing each step of A' as a power series in the x_i's and then truncating the power series for each step such that the maximum degree is quadratic. If a step h_i has a nonzero constant term in the denominator and therefore cannot be expanded into a power series, then one or more of the x_i will be replaced by $x_i - g_i$, $g_i \in G$, until all denominators have nonzero constant terms. This substitution is always possible since G has an infinite number of elements. The system $z_k = \sum_{i=1}^{r} \sum_{j \leq i} g_{ijk}(x_i - g_i)(x_j - g_j)$, $k = 1, 2, \ldots, t$, would then be computed, from which the original system S can be computed without additional m/d steps.

Let L_0, L_1, and L_2 be three linear operators over power series that extract the constant, linear, and quadratic components of the power series and let $L = L_0 + L_1 + L_2$. Each step h_i' of A' will be replaced by a step h_i of A such that $L(h_i') = L(h_i) = h_i$. Since the system S is quadratic, each of the outputs satisfies $z_k = L(z_k)$, showing that the algorithm A will indeed compute the system S. The modifications will replace each m/d step with only one m/d step and non-m/d steps with non-m/d steps, hence $\mu_B(A; G) = \mu_B(S; G)$.

Each of the inputs obviously satisfies $L(x_i - g_i) = x_i - g_i$. We will assume that $L(h_i) = h_i$ $(1 \le i \le k)$, and show how to construct the next element h_{k+1} of A from h_{k+1}' in A'. If $h_{k+1}' = g_1 h_i' + g_2 h_j'$ for $i, j < k$, where g_1 and g_2 are not related to g_i above, then $h_{k+1} = g_1 h_i + g_2 h_j$ where h_i and h_j are the steps in A corresponding to h_i' and h_j' in A'. If $h_{k+1}' = h_i' \cdot h_j'$, then h_{k+1}' is replaced by a sequence of steps that compute

$$L_0(h_{k+1}) = L_0(h_i)L_0(h_j),$$

$$L_1(h_{k+1}) = L_0(h_i)L_1(h_j) + L_1(h_i)L_0(h_j), \text{ and}$$

$$L_2(h_{k+1}) = L_0(h_i)L_2(h_j) + L_1(h_i)L_1(h_j) + L_2(h_i)L_0(h_j).$$

We then compute $h_{k+1} = L_0(h_{k+1}) + L_1(h_{k+1}) + L_2(h_{k+1})$. The only multiplication step in this calculation is that of $L_1(h_i)L_1(h_j)$, since multiplications by $L_0(x)$ are by an element of G. This substitution has therefore not changed the multiplicative complexity of the algorithm.

If $h_{k+1} = h_i \div h_j$, then h_{k+1} is replaced by a sequence of steps that compute

$$L_0(h_{k+1}) = L_0(h_i)/L_0(h_j),$$

$$L_1(h_{k+1}) = (L_1(h_i)L_0(h_j) - L_0(h_i)L_1(h_j))/L_0^2(h_j), \text{ and}$$

$$L_2(h_{k+1}) = (L_2(h_i)L_0^2(h_j) - L_0(h_i)L_2(h_j)L_0(h_j) - L_1(h_i)L_1(h_j)L_0(h_j)$$
$$+ L_0(h_i)L_1^2(h_j))/L_0^3(h_j).$$

We then compute $h_{k+1} = L_0(h_{k+1}) + L_1(h_{k+1}) + L_2(h_{k+1})$. There is again only one multiplication step in this set of calculations, $[L_0(h_i)L_1(h_j) - L_1(h_i)L_0(h_j)]L_1(h_j)$, since both multiplications and divisions by $L_0(x)$ are not counted. ∎

It has now been shown that each step of A' can be replaced by a step that is of at most second degree in the indeterminates, and that the resulting algorithm A is division-free. An algorithm for a system of quadratic forms that has these attributes is called a *quadratic algorithm*. We define a *bilinear algorithm* to be a division-free algorithm for a system of bilinear forms such that each step is of at most second degree in the indeterminates and such that no linear steps include elements from both sets of indeterminates. An example of an algorithm that is quadratic, but not bilinear, is the computation of the system $x_1y_1+x_2y_2$ as $(x_1+y_2)y_1+(x_2-y_1)y_2$. The following example will illustrate the use of Theorem 2.8.

Example 2.3. The following bilinear system has been proposed by Feig [16] as one that may be computed in the minimum number of m/d steps with at least one division step. Let S of Theorem 2.8 be

$$C(x)z = \begin{bmatrix} x_3 & x_4 & 0 & 0 \\ x_0+x_3 & x_1 & 0 & 0 \\ 0 & 0 & x_0 & 0 \\ 0 & 0 & x_2 & x_1 \\ 0 & 0 & 0 & x_0-x_1 \\ 0 & 0 & x_1 & 0 \end{bmatrix} \begin{bmatrix} y_0 \\ y_1 \\ v_0 \\ v_1 \end{bmatrix}$$

and the algorithm A' be defined by the following steps:

$S_1 = x_0+y_1, S_2 = x_1+y_0, S_3 = x_3-y_1, S_4 = x_4+y_0, S_5 = x_0-x_1,$

$P_1 = x_0x_1, P_2 = S_1S_2, P_3 = S_3y_0, P_4 = S_4y_1, P_5 = x_0v_0,$

$S_6 = v_1+P_5, S_7 = x_2-P_1,$

$P_6 = x_1S_6, P_7 = v_0S_7, P_8 = v_1S_5,$

$S_8 = P_3+P_4, S_9 = P_2+P_3, S_{10} = S_9-P_1, S_{11} = P_6+P_7, S_{12} = P_6+P_8,$

$D = S_{12} \div x_0, S_{13} = D-v_1,$

where the outputs are the vector $[S_8\ S_{10}\ P_5\ S_{11}\ P_8\ S_{13}]^T$. The S_i steps are sums, the P_i steps are products, and the D step is a division. Feig [16] has proven that this system requires 9 m/d steps.

A' has two multiplication steps, P_6 and P_7, that are cubic in the indeterminates and one division step, D. The denominator of D must have a nonzero constant term to convert A' into A as in Theorem 2.8. To insure this, x_0 is replaced by $S_0 = x_0+g_0$ where $g_0 \in G$ and $g_0 \neq 0$. For simplicity, let $g_0 = -1$. After modifying the algorithm as in Theorem 2.8, each output that depends on S_0 (steps S_{14}, S_{15}, and S_{16} in A) must be converted back using non-m/d steps to obtain a new division-free algorithm that computes S. The new algorithm computes the system in the same order as A'.

The steps in the algorithm A are

$$S_0 = x_0+1, S_1 = S_0+y_1, S_2 = x_1+y_0, S_3 = x_3-y_1, S_4 = x_4+y_0,$$

$$S_5 = S_0-x_1,$$

$$P_1 = S_0x_1, P_2 = S_1S_2, P_3 = S_3y_0, P_4 = S_4y_1, P_5 = S_0v_0,$$

$$S_6 = v_1+P_5, S_7 = x_2-P_1, S_8 = v_0+v_1, S_9 = x_2-x_1$$

$$P_6 = x_1S_8, P_7 = v_0S_9, P_8 = v_1S_5,$$

$$S_{10} = P_3+P_4, S_{11} = P_2+P_3, S_{12} = S_{11}-P_1, S_{13} = P_6+P_7, S_{14} = P_6+P_8,$$

$$S_{15} = S_{14}-v_1,$$

$$P_9 = x_1v_0,$$

$$S_{16} = S_{15}-P_9, S_{17} = S_{16}+v_1, S_{18} = S_{12}-y_0, S_{19} = P_5-v_0, S_{20} = P_8-v_1,$$

and the output vector is $[S_{10} \ S_{18} \ S_{19} \ S_{13} \ S_{20} \ P_9]^T$.

The only deviation from the procedure described in Theorem 2.8 is that steps that would have been equal to zero were omitted. This occurs when one or more of the linear operators $L_0, L_1,$ or L_2 is equal to zero for an original m/d step that was of order greater than 2. Examination of the algorithm shows that frequently it is unnecessary to replace a step that is later to be decomposed into its constant, linear, and quadratic parts; these components can always be computed directly instead. Thus many extra non-m/d steps are included in the algorithm that are neither outputs nor used in the computation of any of the outputs. In the algorithm A, the steps S_6, $S_7, S_{14}, S_{15}, S_{16},$ and S_{17} are all unnecessary computations.

Since there are only 9 nonzero entries in the Φ matrix of this example, the system could have been computed with two addition steps, followed by nine multiplication steps, and ending with three addition steps, for a total of only five non-m/d steps. Thus the strict application of Theorem 2.8 does not necessarily yield an algorithm with the minimum number of non-m/d steps.

2.4.2. Bilinear Systems and Noncommutative Algorithms

The systems that have been introduced up to this point have been analyzed for a set of indeterminate inputs. No restrictions have yet been placed on the nature of these indeterminate quantities other than those implicit in semilinear systems. Many real systems of interest can be decomposed in such a way that they represent a smaller system with more complicated indeterminates. An example would be a 4×4 matrix multiplication expressed as a 2×2 matrix multiplication where each of the entries is a 2×2 matrix itself. The original system has been expressed as the direct (or tensor) product of two smaller systems, a concept that will be studied in the next section.

The simple example just suggested illustrates a possible problem. If an algorithm is known for 2×2 matrix multiplication, then that algorithm can be iterated to accomplish the 4×4 matrix multiplication. The problem with this procedure is that the algorithm for computing the 2×2 matrix product must have all its multiplication steps in the proper order so that when the indeterminates are replaced by 2×2 matrices each of these steps is computing the correct 2×2 product. Since matrix multiplication is not commutative, we must insure that the algorithm is noncommutative before extending it in this way. This motivates the need to introduce the concept of a *noncommutative algorithm* to extend some of the ideas already discussed to systems defined for other algebraic structures in which multiplication is not commutative.

Definition 2.8. *A noncommutative algorithm is one that does not depend on the commutative law of multiplication. We denote by $\bar{\mu}(S)$ the minimum number of multiplications required by any noncommutative algorithm to compute the system S of bilinear forms.*

Theorem 2.9. *For a bilinear system S, $\mu(S) \leq \bar{\mu}(S) \leq 2\mu(S)$.*

Proof. In Theorem 2.8 it was proved that for any quadratic system an algorithm exists in which each m/d step is of the form $h(i) = M_i(x)N_i(x)$, where $M_i(x)$ and $N_i(x)$ are both linear in the indeterminates. When S is bilinear this means that an algorithm exists in which each m/d step is of the form $h(i) = M_i(x, y)N_i(x, y)$, where $M_i(x, y)$ and $N_i(x, y)$ are both linear in the indeterminates. This algorithm will be commutative unless the coefficients of the y terms are zero in each $M_i(x, y)$ and the coefficients of the x terms are zero in each $N_i(x, y)$, in which case $\mu(S) = \bar{\mu}(S)$. When the algorithm is commutative, it can be converted to a noncommutative algorithm by replacing each m/d step that is not of the form $M_i(x)N_i(y)$ with two m/d steps, $M_i(x, 0)N_i(0, y)$ and $N_i(x, 0)M_i(0, y)$, and adding them together. The quadratic terms that are not bilinear will not be missed since they ultimately cancel to zero in a bilinear system anyway, and all other terms from the original algorithm will remain. In the worst case this algorithm will require twice as many multiplications as a commutative algorithm, hence $\mu(S) \leq \bar{\mu}(S) \leq 2\mu(S)$. ∎

In the proof of Theorem 2.9 the algorithm that results from converting the nonbilinear steps to bilinear steps is not necessarily a minimal noncommutative algorithm. The notion of a noncommutative algorithm is useful in extending the indeterminates used in the definition of a system to include structures other than real or complex numbers, such as matrices. The next theorem states a property of bilinear systems that is analogous to Theorem 2.8 for quadratic systems.

Theorem 2.10. *Let S be a system of bilinear forms. If G has infinitely many elements, then a minimal noncommutative algorithm A computing S exists such that each m/d step of A is of the form $h(i) = M_i(x)N_i(y)$, where $M_i(x)$ and $N_i(y)$ are linear in x and y respectively.*

Proof. Let S be the system $z_k = \sum\limits_{i=1}^{r} \sum\limits_{j=1}^{s} g_{ijk} x_i y_j$, $k = 1, 2, \ldots, t$. Let $A' = (h'_1, h'_2, \ldots, h'_n)$ be a minimal noncommutative algorithm computing S. The substitutions as outlined in Theorem 2.8 do not depend on the commutative law of multiplication, thus if these substitutions are made the resulting algorithm will remain noncommutative. Each step will be at most quadratic in the indeterminates and therefore both multiplicands in every product must be linear in the indeterminates. ∎

Another useful technique for analyzing systems of bilinear forms is the construction of a trilinear (tensor) system from a system of bilinear forms. Given the bilinear system S whose outputs are

$$f_k = \sum_{i=1}^{r} \sum_{j=1}^{s} g_{ijk} x_i y_j, \quad k = 1, 2, \ldots, t,$$

we multiply each output by an indeterminate, z_k, and sum to form the tensor

$$T = \sum_{i=1}^{r} \sum_{j=1}^{s} \sum_{k=1}^{t} g_{ijk} x_i y_j z_k. \tag{2.1}$$

If $\bar{\mu}(S) = n$, then Theorem 2.1 guarantees that in a minimal noncommutative bilinear algorithm computing S, where the multiplication steps are $m_l = (\sum\limits_{i=1}^{r} \alpha_{il} x_i)(\sum\limits_{j=1}^{s} \beta_{jl} y_j)$, $l = 1, 2, \ldots, n$, $\alpha_{il}, \beta_{jl} \in G$, each of the outputs can be expressed as $f_k = \sum\limits_{l=1}^{n} \gamma_{kl} m_l$, $\gamma_{kl} \in G$. We substitute this into (2.1) and rearrange the sums to obtain

$$T = \sum_{l=1}^{n} m_l (\sum_{k=1}^{t} \gamma_{kl} z_k) = \sum_{l=1}^{n} (\sum_{i=1}^{r} \alpha_{il} x_i)(\sum_{j=1}^{s} \beta_{jl} y_j)(\sum_{k=1}^{t} \gamma_{kl} z_k).$$

From this characterization it is seen that $\bar{\mu}(S)$ is the rank of the tensor T. Two other distinct systems can be constructed from S with the same noncommutative multiplicative complexity as S by grouping all terms multiplying either the x_i's or y_j's. This procedure is called transposing the tensor and is particularly useful when an algorithm is already known for a transpose of a system under consideration. Another application arises when a system is *equivalent* to one of its transposes, a property possessed by systems of cyclic convolution. When this occurs one set of constants (α's, β's, or γ's) is often more complicated (requires more additions) than the others, in which case it may be possible to transpose an algorithm so that the more complicated coefficients occur in a part of the algorithm that may be precomputed. The following example, originally presented by Winograd [44], demonstrates this situation.

Example 2.4. Let S be the system

$$\begin{bmatrix} z_0 \\ z_1 \\ z_2 \end{bmatrix} = \begin{bmatrix} x_0 & x_1 & x_2 \\ x_1 & x_2 & x_0 \\ x_2 & x_0 & x_1 \end{bmatrix} \begin{bmatrix} y_0 \\ y_1 \\ y_2 \end{bmatrix}$$

which is a cyclic convolution of length 3. Using techniques that will be developed in the next chapter, the following algorithm A' can be derived.

$$m_0' = \frac{(x_0+x_1+x_2)}{3}(y_0+y_1+y_2)$$

$$m_1' = \frac{(x_0-x_2)}{3}(y_0-y_1)$$

$$m_2' = \frac{(x_1-x_2)}{3}(y_2-y_1)$$

$$m_3' = \frac{(x_0-x_1)}{3}(y_0-y_2)$$

$$z_0 = m_0'+m_1'-2m_2'+m_3'$$

$$z_1 = m_0'+m_1'+m_2'-2m_3'$$

$$z_2 = m_0'-2m_1'+m_2'+m_3'$$

We form the tensor

$$T = z_0x_0y_0+z_0x_1y_1+z_0x_2y_2$$

$$+z_1x_1y_0+z_1x_2y_1+z_1x_0y_2$$

$$+z_2x_2y_0+z_2x_0y_1+z_2x_1y_2.$$

Regrouping terms yields

$$T = x_0(z_0y_0+z_1y_2+z_2y_1)$$

$$+x_1(z_1y_0+z_2y_2+z_0y_1)$$

$$+x_2(z_2y_0+z_0y_2+z_1y_1)$$

which is the tensor representation of the system

$$\begin{bmatrix} x_0 \\ x_1 \\ x_2 \end{bmatrix} = \begin{bmatrix} z_0 & z_1 & z_2 \\ z_1 & z_2 & z_0 \\ z_2 & z_0 & z_1 \end{bmatrix} \begin{bmatrix} y_0 \\ y_2 \\ y_1 \end{bmatrix} \qquad (2.2)$$

which is identical to S, with the x_i's and z_i's swapped and y_1 swapped with y_2. The algorithm A' yields the representation of the tensor T as

$$T = \frac{(x_0+x_1+x_2)}{3}(y_0+y_1+y_2)(z_0+z_1+z_2)$$

$$+\frac{(x_0-x_2)}{3}(y_0-y_1)(z_0+z_1-2z_2)$$

$$+\frac{(x_1-x_2)}{3}(y_2-y_1)(-2z_0+z_1+z_2)$$

$$+\frac{(x_0-x_1)}{3}(y_0-y_2)(z_0-2z_1+z_2).$$

Regrouping the x_i terms yields the following algorithm for the system of (2.2).

$$m_0 = \frac{(z_0+z_1+z_2)}{3}(y_0+y_1+y_2)$$

$$m_1 = \frac{(z_0+z_1-2z_2)}{3}(y_0-y_1)$$

$$m_2 = \frac{(-2z_0+z_1+z_2)}{3}(y_2-y_1)$$

$$m_3 = \frac{(z_0-2z_1+z_2)}{3}(y_0-y_2)$$

$$x_0 = m_0+m_1+m_3$$

$$x_1 = m_0+m_2-m_3$$

$$x_2 = m_0-m_1-m_2$$

Since the system of (2.2) is equivalent to S, this algorithm is in turn equivalent to an algorithm A computing S, where A is obtained by swapping each x_i and z_i and swapping y_1 and y_2. The algorithm A is

$$m_0 = \frac{(x_0+x_1+x_2)}{3}(y_0+y_1+y_2)$$

$$m_1 = \frac{(x_0+x_1-2x_2)}{3}(y_0-y_2)$$

$$m_2 = \frac{(-2x_0+x_1+x_2)}{3}(y_1-y_2)$$

$$m_3 = \frac{(x_0-2x_1+x_2)}{3}(y_0-y_1)$$

$$z_0 = m_0+m_1+m_3$$

$$z_1 = m_0+m_2-m_3$$

$$z_2 = m_0-m_1-m_2$$

Assuming that S is a semilinear system with the x_i's known (i.e., we are cyclically convolving with a fixed kernel), then any function of the x_i's alone may be precomputed. The algorithm A requires 5 pre-multiply additions and 6 post-multiply additions for a total of 11 additions. A', on the other hand, has the same set of pre-multiply additions, but has more complicated post-multiply coefficients. The most efficient way of doing these post-multiply additions is to compute

$$a_1 = m'_1-m'_2, \ a_2 = m'_2-m'_3, \ a_3 = m'_3-m'_1,$$

$$a_4 = a_1-a_2, \ a_5 = a_2-a_3, \ a_6 = a_3-a_1,$$

$$z_0 = m'_0+a_4, \ z_1 = m'_0+a_5, \ z_2 = m'_0+a_6$$

for a total of 9 post-multiply additions. Therefore, transposing the tensor has reduced the total number of additions from 14 to 11 by moving the most complicated coefficients to the part of the computation that can be precomputed.

The notion of *equivalence* of systems is an important concept and deserving of a formal definition.

Definition 2.9. *Let S be the system of t bilinear forms $z_k = \sum_{i=1}^{r} \sum_{j=1}^{s} g_{ijk} x_i y_j$, $k = 1, 2, \ldots, t$, and let S' be the system $z'_k = \sum_{i=1}^{r} \sum_{j=1}^{s} g'_{ijk} x'_i y'_j$, $k = 1, 2, \ldots, t$. S and S' are equivalent if three nonsingular matrices A, B, and C with entries $a_{i,i'}$ $(1 \le i, i' \le r)$, $b_{j,j'}$ $(1 \le j, j' \le s)$, and $c_{k,k'}$ $(1 \le k, k' \le t)$, respectively, all in G exist such that*

$$\sum_{k=1}^{t} c_{k,k'} z_k = \sum_{i'=1}^{r} \sum_{j'=1}^{s} g'_{i'j'k'} \left[\sum_{i=1}^{r} a_{i,i'} x_i \right] \left[\sum_{j=1}^{s} b_{j,j'} y_j \right], \quad k' = 1, 2, \ldots, t.$$

Thus two bilinear systems are equivalent if the indeterminates (and outputs) of one system can be replaced by linear combinations (over G) of the indeterminates (and outputs) of the other system to yield the second system itself, and vice versa. The simplest equivalences occur when the coefficients (g_{ijk}'s) of two systems are identical except for possible renumbering of the i, j, and k subscripts, in which case A, B, and C are just permutation matrices. A more detailed discussion of equivalence is presented in the next chapter.

2.4.3. Direct Products and Direct Sums of Systems

Two methods will now be introduced for constructing algorithms for large systems from algorithms for smaller systems. These consist of the tensor (or direct or Kronecker) product method and the direct sum method.

Definition 2.10. *Let S be the system of bilinear forms $\Phi(x)y$ and S' the system $\Phi'(x')y'$. The tensor product $S'' = S \otimes S'$ is the system $\Phi''(x'')y''$, where $\Phi''(x'') = \Phi(x) \otimes \Phi'(x)$ is the direct product of $\Phi(x)$ and $\Phi'(x')$, being the matrix formed by replacing each entry ϕ_{ij} of Φ with the block $\phi_{ij}\Phi'$. If Φ is $t \times n$ and Φ' is $t' \times n'$, then Φ'' will be a $tt' \times nn'$ matrix.*

Suppose A is a noncommutative algorithm computing S and A' is a noncommutative algorithm computing S'. The algorithm A'' computing S'' is formed by replacing each m/d step of A with an application of A'. A'' computes S'' using $\mu(A)\mu(A')$ m/d steps through the tensor product construction. If A and A' had been minimal algorithms, then A'' generally would not be a minimal algorithm, although it could possibly be minimal.

Example 2.5. Let S be the system

$$\begin{bmatrix} z_0 \\ z_1 \end{bmatrix} = \begin{bmatrix} x_0 & x_1 \\ x_1 & x_0 \end{bmatrix} \begin{bmatrix} y_0 \\ y_1 \end{bmatrix}$$

and let S' be the system

$$\begin{bmatrix} z'_0 \\ z'_1 \end{bmatrix} = \begin{bmatrix} x'_0 & -x'_1 \\ x'_1 & x'_0 \end{bmatrix} \begin{bmatrix} y'_0 \\ y'_1 \end{bmatrix}. \tag{2.3}$$

S is a system that computes a length-2 cyclic convolution and S' is a system that computes the complex product $(x'_0 + jx'_1)(y'_0 + j'_1) = (x'_0 y'_0 - x'_1 y'_1) + j(x'_1 y'_0 + x'_0 y'_1)$. The direct product $S'' = S \otimes S'$ is a length-2 cyclic convolution of complex numbers and is represented by

$$\begin{bmatrix} z_0'' \\ z_1'' \\ z_2'' \\ z_3'' \end{bmatrix} = \begin{bmatrix} x_0'' & -x_1'' & x_2'' & -x_3'' \\ x_1'' & x_0'' & x_3'' & x_2'' \\ x_2'' & -x_3'' & x_0'' & -x_1'' \\ x_3'' & x_2'' & x_1'' & x_0'' \end{bmatrix} \begin{bmatrix} y_0'' \\ y_1'' \\ y_2'' \\ y_3'' \end{bmatrix}.$$

This system could be computed using an algorithm for length-2 cyclic convolution in which each multiplication is replaced by a complex product algorithm. Alternatively, since the Φ matrix of the system $S' \otimes S$ is simply a rearrangement of the rows and columns of the Φ matrix for the system $S \otimes S'$, S'' could also be computed by a complex product algorithm in which each multiplication is replaced by an algorithm for length-2 cyclic convolution.

Definition 2.11. *Let S be the system of bilinear forms $\Phi(x)y$ and S' the system $\Phi'(x')y'$. We define the direct sum $S'' = S \oplus S'$ as the system $\Phi''(x'')y''$ where $\Phi''(x'')$ is the $t+t' \times n+n'$ block diagonal matrix in which the upper left block is the $t \times n$ matrix $\Phi(x)$ and the lower right block is the $t' \times n'$ matrix $\Phi'(x')$. The vector y'' is the concatenation of the entries of y and y'.*

It is conjectured that $\mu_B(S''; G) = \mu_B(S; G) + \mu_B(S'; G)$, but the conjecture has not been proven in general. It is also conjectured that all minimal algorithms for computing S'' compute S and S' separately. The combination of these two conjectures is referred to as the *Direct Sum Conjecture* [17, 47] and is still an open problem.

Direct sums of systems occur when two or more distinct systems must be computed. If any of the indeterminates are shared between two systems computed simultaneously, then this will be called the *union* of the two systems. A direct sum is a special type of union of systems.

For example, the computation of $x_0 y_0$ and $x_0 y_1$ is the union of $x_0 y_0$ with $x_0 y_1$. On the other hand, the union of $x_0 y_0$ and $x_1 y_1$ is also a direct sum of $x_0 y_0$ and $x_1 y_1$. Both of these systems require two m/d steps, but the first system can be computed as $x_0 y_0$ and $x_0 y_0 + x_0 (y_1 - y_0)$, allowing one result to be used in computing the other, which is not possible in the second system because of the direct sum conjecture (which although unproven in general can be shown to hold in this case).

2.5. Summary of Chapter 2

An algebraic framework has been established for the analysis of the multiplicative complexity of computational systems, and in particular for bilinear systems. Semilinear systems were introduced to facilitate analysis of some common systems for which one set of the system inputs are known in advance, allowing certain functions of these inputs to be precomputed.

The *row-rank theorem* (Theorem 2.4) and the *column-rank theorem* (Theorem 2.5) provide lower bounds on the multiplicative complexity of many systems. These theorems, together with the specializations guaranteed by Theorems 2.6 and 2.7, will be the principal tools needed to determine the multiplicative complexity of polynomial products modulo other polynomials, and ultimately the multiplicative complexity of the one-dimensional discrete Fourier transform.

Several other ideas have been presented in this chapter that are more useful in computing a given system than in evaluating the multiplicative complexity. These include the concepts of direct products and direct sums of systems, noncommutative algorithms, and the trilinear formulation of bilinear systems. The conversion of division steps into multiplication steps as suggested by Theorem 2.8 may also be helpful in this regard.

It will be discovered in the succeeding chapters that sometimes, particularly for large systems, algorithms developed to minimize the number of multiplications and divisions are impractical to implement because of the limited arithmetic precision of practical hardware processors. Although such algorithms are mathematically sound, they typically use very large numbers from the field of constants G. This causes the algorithm to be less numerically stable and generates products with large elements of G that are classified as non-m/d steps, but should realistically be counted as multiplications since the alternative is to implement them with an excessively large number of additions and subtractions. Algorithms that are derived through the theory of multiplicative complexity are not always practical to implement, but can serve the purpose of defining a lower bound on the number of multiplications necessary to compute a system, which is useful in evaluating more practical algorithms.

CHAPTER 3

Convolution and Polynomial Multiplication

Convolution, or digital filtering, is one of the most common operations used in modern signal processing. Aperiodic convolution can be expressed as a product of polynomials, and cyclic convolution, commonly used in block filtering techniques, is equivalent to a product in a polynomial ring. This chapter begins by applying the theory developed in Chapter 2 to the analysis of the multiplicative complexity of polynomial products, which can be reformulated as convolutions through these equivalences.

The multiplicative complexity of products of polynomials with indeterminate coefficients is first determined. This analysis is generalized to systems that are the union of products of a single polynomial with several other polynomials. Direct products of systems of polynomial multiplication are shown to be equivalent to direct sums of similar systems. This equivalence is used to derive expressions for the multiplicative complexity of multidimensional cyclic convolution for any dimensionality and lengths.

3.1. Aperiodic Convolution / Polynomial Multiplication

Let $x = [x_0 \, x_1 \, \cdots \, x_m]^T$ and $y = [y_0 \, y_1 \, \cdots \, y_n]^T$ be two vectors of indeterminates. The *aperiodic (or linear) convolution* $x * y$ of x and y is defined by

$$z_j = \sum_{i=\max(0,\,j-n)}^{\min(j,\,m)} x_i y_{j-i}, \quad j = 0, 1, \ldots, m+n, \tag{3.1}$$

where the max and min functions give the maximum or minimum, respectively, of the enclosed arguments. Sometimes, for notational convenience, it is assumed that y_j is zero for $j < 0$ and $j > n$, in which case the summation over i in (3.1) could be from 0 to m.

The outputs of this convolution are the coefficients of the polynomial $z(u) = \sum_{i=0}^{m+n} z_i u^i$, which is the product of $x(u) = \sum_{i=0}^{m} x_i u^i$ and $y(u) = \sum_{i=0}^{n} y_i u^i$. Therefore the system of (3.1) is equivalent to a product of polynomials of degree m and n and will be denoted $PM(m,n)$. The base set for this system is $B = G \cup \{x_0, x_1, \ldots, x_m\} \cup \{y_0, y_1, \ldots, y_n\}$. The results of Chapter 2 can be applied to this system to obtain a lower bound on the number of required m/d steps.

Lemma 3.1. $\mu_B(PM(m, n); G) \geq m+n+1.$

Proof. The system $PM(m, n)$ can be represented as the product Φy where Φ is an $(m+n+1) \times (n+1)$ matrix whose entries are the indeterminates $\phi_{ij} = x_{i-j}$ $(0 \leq i-j \leq m)$ and zero otherwise, and $y = [y_0 \, y_1 \, \cdots \, y_n]^T$ is a vector of $n+1$ indeterminates. Since all the entries of Φ are indeterminate elements of H, no nontrivial linear combination (over G) of the $m+n$ distinct entries of Φ can yield an element of G. From this we conclude that no nontrivial linear combination (over G) of the rows of Φ yields an element of G^{n+1}. Therefore $\rho_r(\Phi) = m+n+1$ and by Theorem 2.4 we obtain $\mu_B(PM(m, n); G) \geq m+n+1.$ ∎

We will now demonstrate an algorithm that computes $PM(m, n)$ using $m+n+1$ m/d steps, showing that $\mu_B(PM(m, n); G) = m+n+1$. This algorithm is known as the Toom-Cook or Karatsuba-Toom algorithm and is an application of the Lagrange interpolation formula to the product polynomial at $m+n+1$ distinct points in G. The Lagrange interpolation formula computes the coefficients of an Nth degree polynomial from the functional value of the polynomial at $N+1$ distinct points. For our application, we evaluate $x(g_i)$ and $y(g_i)$, $g_i \in G$, at $m+n+1$ distinct elements of G, compute $z(g_i) = x(g_i)y(g_i)$ at each of these points, and then use Lagrangian interpolation to compute the coefficients of $z(u)$. Algorithms of this type are sometimes called *evaluation-interpolation* algorithms because of this procedure.

The Toom-Cook algorithm can also be formulated as an application of the polynomial Chinese remainder theorem (CRT) in which polynomial products are carried out by reducing both the multiplicand and multiplier modulo $m+n+1$ distinct linear polynomials with coefficients in G, performing $m+n+1$ scalar multiplications of the residues, and then reconstructing the product polynomial from the residues. The CRT formulation generalizes to irreducible modulus polynomials of degree greater than one, resulting in near-minimal algorithms that may require significantly fewer non-m/d steps than minimal algorithms.

The *Chinese remainder theorem* for polynomials states that a polynomial $X(u)$ of degree n can be uniquely reconstructed from its residues modulo a set of relatively prime polynomials whose product is of degree $n+1$ or greater. We will assume that each of the modulus polynomials is monic; if not we could divide through by the leading coefficient without affecting the statement of the theorem. Let $P_i(u) = u^{n_i} + \sum_{j=0}^{n_i-1} g_{ij} u^j$, $i = 1, 2, \ldots, s$, where $(P_i(u), P_j(u)) = 1$ for $i \neq j$, $n_i = \deg P_i(u)$, and $\sum_{i=1}^{s} n_i \geq n$. Let $X_i(u)$ denote the residue $X(u) \pmod{P_i(u)}$, then the polynomial $X(u)$ may be reconstructed from its residues as

$$X(u) \equiv \sum_{i=1}^{s} X_i(u) \frac{P(u)}{P_i(u)} Q_i(u) \quad (\mathrm{mod}\, P(u)) \tag{3.2}$$

where $P(u) = \prod_{i=1}^{s} P_i(u)$ and each $Q_i(u)$ satisfies $Q_i(u)P(u)/P_i(u) \equiv 1 \,(\mathrm{mod}\, P_i(u))$. When a modulus polynomial $P_i(u)$ is linear, the corresponding $Q_i(u)$ is constant and equal to $1/\prod_{j \neq i} P_j(-g_{i0})$. In general, if $\deg P_i = n_i$, then $\deg Q_i = n_i - 1$ and $Q_i(u) \equiv (\prod_{j \neq i} P_j(u))^{-1} \,(\mathrm{mod}\, P_i(u))$.

We will now apply the CRT to develop a minimal algorithm for multiplication of polynomials. Let $P(u) = \prod_{i=1}^{m+n+1} P_i(u)$ where $P_i(u) = u - g_i$, $g_i \in G$, $i = 1, 2, \ldots,$ $m+n+1$, and $g_i \neq g_j$ for $i \neq j$. The polynomials to be multiplied are $x(u) = \sum_{i=0}^{m} x_i u^i$ and $y(u) = \sum_{i=0}^{n} y_i u^i$. Let $x_{r_i}(u)$ denote the residue $x(u) \,(\mathrm{mod}\, P_i(u))$ and $y_{r_i}(u)$ the residue $y(u) \,(\mathrm{mod}\, P_i(u))$. Residue reduction yields the polynomial of lowest degree congruent to the given polynomial modulo the modulus polynomial (this polynomial will always have degree less than the modulus polynomial). Residue reduction modulo a monic linear polynomial is simply the evaluation of the input polynomial at the root of the modulus polynomial, therefore we have

$$x_{r_i}(u) = \sum_{j=0}^{m} x_j g_i^j, \quad i = 0, 1, \ldots, m+n+1$$

and

$$y_{r_i}(u) = \sum_{j=0}^{n} y_j g_i^j, \quad i = 0, 1, \ldots, m+n+1.$$

These computations require no m/d steps since all multiplications are by elements of G.

Then $m+n+1$ m/d steps are executed to obtain the residues $z_{r_i}(u) = x_{r_i}(u)y_{r_i}(u)$, $i = 1, 2, \ldots, m+n+1$. The reconstruction polynomials, $Q_i(u) = 1/\prod_{j \neq i} P_j(g_i)$ are constant and in G, and $P(u)/P_i(u)$ is an $m+n^{th}$ degree polynomial with coefficients in G. The coefficients of the z_{r_i}'s in the reconstruction are therefore all elements of G, hence the reconstruction requires no m/d steps. The system $PM(m,n)$ has been computed using $m+n+1$ m/d steps, which, together with Lemma 3.1, proves the following theorem.

Theorem 3.1. *If G contains at least $m+n$ distinct elements then $\mu_B(PM(m,n); G) = m+n+1$.*

The requirement on the number of distinct elements of G is one less than might have been expected because one m/d step could have been $x_m y_n$, yielding the CRT reconstruction

$$z(u) = z'(u) + x_m y_n P(u)$$

where

$$z'(u) = \sum_{i=1}^{m+n} z_{r_i}(u) \frac{P(u)}{P_i(u)} Q_i(u) \ (\mathrm{mod} \ P(u))$$

and the polynomial $P(u)$ is composed of $m+n$ distinct monic linear polynomial factors with coefficients in G. This identity is referred to by Winograd [47] as using the modulus polynomial $u-\infty$, since the relation between the polynomial coefficients in $z(u) = x(u)y(u)$ is equivalent to that in $u^{m+n} z(1/u) = u^m x(1/u) u^n y(1/u)$. The modulus polynomial u in the second representation is $1/u$ in the first, implying $1/u = 0$ or $u = \infty$. Use of this identity is also known as interpolation at $1/u$ [27] for obvious reasons. The following example shows how this identity is used as part of the Toom-Cook algorithm.

Example 3.1. Let S be the system $z(u) = x(u)y(u)$, where $x(u) = x_0 + x_1 u$, $y(u) = y_0 + y_1 u + y_2 u^2$, and $z(u) = z_0 + z_1 u + z_2 u^2 + z_3 u^3$, which is the system $PM(1,2)$. Theorem 3.1 states that 4 is the least number of multiplications that can be used to compute $PM(1,2)$. If $0, 1, -1$, and ∞ are chosen as the rational roots of the modulus polynomials in the CRT, then the following 4 multiplications are a basis over G for the output coefficients.

$$m_1 = x_0 y_0$$
$$m_2 = (x_0 + x_1)(y_0 + y_1 + y_2)/2$$
$$m_3 = (x_0 - x_1)(y_0 - y_1 + y_2)/2$$
$$m_4 = x_1 y_2.$$

The CRT reconstruction is

$$z_0 = m_1$$
$$z_1 = m_2 - m_3 - m_4$$
$$z_2 = -m_1 + m_2 + m_3$$
$$z_3 = m_4.$$

Winograd [47] has shown that the *only* minimal algorithms for polynomial multiplication are those that use the CRT as described in the remarks leading to Theorem 3.1, or those that use the CRT and the $u-\infty$ modulus.

3.2. Polynomial Multiplication Modulo an Irreducible Polynomial

The results of §3.1 can be extended to polynomial products modulo polynomials with coefficients in G. We define the modular polynomial multiplication of $x(u)$ and $y(u)$ modulo $P(u)$, denoted by $MPM(P)$, as

$$z(u) = x(u)y(u) \ (\text{mod } P(u)) \tag{3.3}$$

where $x(u) = \sum_{i=0}^{n-1} x_i u^i$, $y(u) = \sum_{i=0}^{n-1} y_i u^i$, $z(u) = \sum_{i=0}^{n-1} z_i u^i$, the x_i's and y_i's are indeterminates, and $P(u) = u^n + \sum_{i=0}^{n-1} g_i u^i$.

In this definition we always assume that the polynomials to be multiplied are of degree one less than the modulus polynomial. If their degree were larger, then they could be reduced modulo $P(u)$ with no m/d steps, resulting in an equivalent system. If they were of smaller degree, then the system could be analyzed in a similar way, yielding a multiplicative complexity that might be less than that of the complete system.

A polynomial is *irreducible over a field F* if it cannot be factored into polynomials of lesser degree in $F[u]$. Irreducible polynomials are analogous to prime numbers in the ring of integers. We will first consider modular polynomial multiplications where the modulus polynomial P is either irreducible over G, or a power of an irreducible polynomial.

Clearly any system of the type (3.3), even those in which $P(u)$ is not irreducible, satisfies $\mu_B(MPM(P); G) \leq 2n-1$ since the input polynomials could be multiplied using the Toom-Cook algorithm in $2n-1$ m/d steps, and then reduced modulo $P(u)$. We will now prove that when $P(u)$ is an irreducible polynomial or power of an irreducible polynomial (over G), exactly $2n-1$ m/d steps are necessary to compute $MPM(P)$. The original proof is due to Winograd [43].

Theorem 3.2. *Let $Q(u)$ be a monic irreducible polynomial over the field G and let n be the degree of $P(u) = Q^k(u)$. If G contains at least $2n-2$ distinct elements then $\mu_B(MPM(P); G) = \bar{\mu}_B(MPM(P); G) = 2n-1$.*

Proof. Let

$$C_P = \begin{bmatrix} 0 & 0 & 0 & \cdots & 0 & -g_0 \\ 1 & 0 & 0 & \cdots & 0 & -g_1 \\ 0 & 1 & 0 & \cdots & 0 & -g_2 \\ \vdots & \vdots & \vdots & \ddots & \vdots & \vdots \\ 0 & 0 & 0 & \cdots & 1 & -g_{n-1} \end{bmatrix}$$

be the (right) companion matrix of P, and let $V_P = \{v \in G^n \mid \exists q \neq 0, \deg q < n,$

$vq(C_P) = 0\}$. Each root of P is an eigenvalue λ of C_P and must satisfy $q(C_P)f = q(\lambda)f$ for any polynomial q, where f is the eigenvector of C_P associated with the eigenvalue λ. When $v \neq 0$ the condition $vq(C_P) = 0$ implies that $q(C_P)$ is singular, and hence that at least one eigenvalue λ of $q(C_P)$ is zero. It follows that $q(\lambda) = 0$, which, since Q is irreducible and thus the minimal polynomial (over G) of λ, implies that $q = Q^r s$, $0 < r < k$, where s is a polynomial with coefficients in G, and $(s, Q) = 1$ ($r < k$ follows from $\deg q < n$). Since s is relatively prime to Q, and Q is the minimal polynomial for each of its roots, then s shares no root with Q, implying that $s(C_P)$ is nonsingular.

Therefore $0 = vQ^r(C_P)s(C_P) = vQ^r(C_P)$ and also $vQ^{k-1}(C_P) = 0$. Since this statement is true for all v in V_P, we can now specify $V_P = \{v \in G^n \mid vQ^{k-1}(C_P) = 0\}$. Therefore V_P is a subspace of G^n with $\dim V_P < n$. When $k = 1$, no polynomial q exists that is divisible by Q and is of degree less than n, thus V_P is the empty set and again $\dim V_P < n$.

Let $z = A(x)y$, then

$$A(x) = [x \mid C_P x \mid C_P^2 x \mid \cdots \mid C_P^{n-1} x]$$

follows from the observation that the coefficients of the polynomial $u \sum_{i=0}^{n-1} a_i u^i$ $(\bmod P(u))$ are $C_P a$ where $a = [a_0\, a_1\, \cdots\, a_{n-1}]^T$. Let t be the minimum number of multiplications needed to compute $MPM(P)$, then $A(x)y = Um$ where U is an $n \times t$ matrix with entries in G and $m = [m_1\, m_2\, \cdots\, m_t]^T$. For all nonzero row vectors $w \in G^n$, $wA(x) \neq 0$, therefore rank $U = n$ and U has n linearly independent columns. The columns of U may be permuted if necessary such that the first n columns are linearly independent. A non-singular $n \times n$ matrix W exists such that $WU = (I \mid U')$ and at least one of its rows is not in V_P since the rows of W span G^n. Permute the rows of W and columns of U such that this row, denoted by w, is the first row. We obtain $wA(x)y = [1\, 0\, \cdots\, 0\, u_1'\, u_2'\, \cdots\, u_{t-n}']m$, therefore $wA(x)y$ can be computed using at most $t-n+1$ multiplications.

We now show that no nontrivial linear combination of the columns of $wA(x)$ is identically zero. Assume

$$0 = \sum_{i=0}^{n-1} wC_P^i x \alpha_i = w(\sum_{i=0}^{n-1} \alpha_i C_P^i)x,$$

then

$$w \sum_{i=0}^{n-1} \alpha_i C_P^i = 0,$$

but $w \notin V_P$ and therefore $\alpha_i = 0$, $i = 0, 1, \ldots, n-1$. The column rank of $wA(x)$ is n, therefore at least n multiplications are required to compute $wA(x)y$ by Theorem 2.5. We already found that $\mu(wA(x)y) \leq t-n+1$, therefore $t-n+1 \geq n$ and $t \geq 2n-1$. Theorem 3.1 states that the product of two $n-1^{st}$ degree polynomials can be computed using only $2n-1$ multiplications, therefore $\mu_B(MPM(P); G) = 2n-1$ for P a power of an irreducible polynomial over G. ∎

Example 3.2. Let S be a system of polynomial multiplication modulo u^2+1, which is equivalent to a product of complex numbers. Since u^2+1 is irreducible over Q, the field of rational numbers, Theorem 3.2 states that $\mu_B(MPM(u^2+1); Q) = 3$. This system is

$$z_0+z_1u = (x_0+x_1u)(y_0+y_1u) \ (\mathrm{mod}\ u^2+1).$$

The companion matrix is $C_{u^2+1} = \begin{bmatrix} 0 & -1 \\ 1 & 0 \end{bmatrix}$. S is expressed in matrix form as

$A(x)y = \begin{bmatrix} x_0 & -x_1 \\ x_1 & x_0 \end{bmatrix}\begin{bmatrix} y_0 \\ y_1 \end{bmatrix}$. Using the Toom-Cook algorithm, with constants 0, ∞, and 1, yields

$$A(x)y = Um = \begin{bmatrix} 1 & -1 & 0 \\ -1 & -1 & 1 \end{bmatrix}\begin{bmatrix} x_0y_0 \\ x_1y_1 \\ (x_0+x_1)(y_0+y_1) \end{bmatrix}$$

as an algorithm that computes S in 3 real multiplications.

Different choices of the constants in the Toom-Cook algorithm will generally affect the number of non-m/d steps needed to compute a system. Investigating transposes of systems may expose other algorithms that need fewer non-m/d steps, particularly for semilinear systems. The complex product algorithm can be transposed into the algorithm

$$\begin{bmatrix} z_0 \\ z_1 \end{bmatrix} = \begin{bmatrix} 1 & -1 & 0 \\ 1 & 0 & 1 \end{bmatrix}\begin{bmatrix} x_0(y_0+y_1) \\ (x_0+x_1)y_1 \\ (x_1-x_0)y_0 \end{bmatrix}$$

which requires 5 additions, but could be done in only 3 additions if x_0 and x_1 were known in advance, permitting x_0+x_1 and x_0-x_1 to be precomputed. This algorithm cannot be directly obtained from the Toom-Cook algorithm, but is a minimal algorithm nonetheless.

Winograd [46] has shown that any minimal algorithm for $MPM(P)$, with P irreducible, can be derived in one of three ways. The first method is to use the Toom-Cook algorithm on the polynomial product and then reduce modulo P. The second method uses three auxiliary polynomials $A(u)$, $B(u)$, and $C(u)$ of degree less than $\deg P$, and with coefficients in G, which satisfy $A(u)B(u)C(u) \equiv 1 \pmod{P(u)}$. Form the products $A(u)x(u)$ and $B(u)y(u)$, reduce each modulo $P(u)$, multiply them using the first method, multiply the result by $C(u)$, and reduce modulo $P(u)$. The last method chooses a field isomorphic to $G[u]/\langle P(u)\rangle$, converts the coefficients to this field, multiplies in the isomorphic field using the first method, and then applies the inverse isomorphism.

Instead of transposing to obtain the second algorithm in Example 3.2, we could have used the second method outlined above with $A(u) = 1$, $B(u) = u+1$, and $C(u) = -(u-1)/2$.

3.3. Polynomial Multiplication Modulo a General Polynomial

This section extends the results of the previous section to modulus polynomials with more than one distinct irreducible factor. Let

$$P(u) = \prod_{i=1}^{k} P_i^{e_i}(u)$$

where the P_i's are distinct monic irreducible polynomials over G. It is possible to reduce the input polynomials modulo each factor $P_i^{e_i}(u)$, multiply corresponding residues, and reconstruct the product modulo $P(u)$ from the residues via the CRT. The only multiplications in this procedure are in the middle part that requires $2e_i \deg P_i - 1$ multiplications to obtain the residue of the product modulo $P_i^{e_i}(u)$. The total number of multiplications is thus $\sum_{i=1}^{k} (2e_i \deg P_i - 1) = 2n - k$, where $n = \deg P$. Therefore $\mu_B(MPM(P); G) \le 2n - k$, where $P(u) \in G[u]$ is a polynomial of degree n with k distinct irreducible factors.

The structure of the proposed algorithm must be carefully examined to show that this number of multiplications is necessary, as well as sufficient. The first part of the algorithm decomposes the system into the direct sum of k disjoint systems. Since the final part has no m/d steps, the overall system is equivalent (per Definition 2.9) to the direct sum of systems computed in the middle part of the algorithm. The following proof, due to Winograd, proves the first half of the Direct Sum Conjecture for this system (the second half is also proved by Winograd, but omitted here).

Theorem 3.3. [43] *Let* $Q_i(u)$, $i = 1, 2, \ldots, k$, *be distinct monic irreducible polynomials over the field* G *and let* n *be the degree of* $P(u) = \prod_{i=1}^{k} Q_i^{e_i}(u)$. *If* G *contains at least* $2r-2$ *distinct elements,* $r = \max_i (e_i \cdot \deg Q_i)$, *then* $\mu_B(MPM(P); G) =$

$\bar{\mu}_B(MPM(P); G) = 2n-k$.

Proof. This proof will use the same basic ideas that were used in proving Theorem 3.2. Let $A(x)y$ be the system consisting of the direct sum of all the residue products as described in the preceding comments. Thus

$$A(x)y = \begin{bmatrix} A_1(x_1) & 0 & \cdots & 0 \\ 0 & A_2(x_2) & \cdots & 0 \\ \vdots & \vdots & \ddots & \vdots \\ 0 & 0 & \cdots & A_k(x_k) \end{bmatrix} \begin{bmatrix} y_1 \\ y_2 \\ \vdots \\ y_k \end{bmatrix}$$

where

$$A_i(x_i) = [x_i \,|\, C_{P_i} x_i \,|\, C_{P_i}^2 x_i \,|\, \cdots \,|\, C_{P_i}^{n_i-1} x_i].$$

and $P_i(u) = Q_i^{e_i}(u)$ for each $i = 1, 2, \ldots, k$. An $n \times t$ matrix U (where t is the minimum number of multiplications) with entries in G exists such that $A(x)y = Um$ where $m = [m_1 \, m_2 \, \cdots \, m_t]^T$. The n rows of $A(x)$ are linearly independent, thus rank $U = n$. Permute, if necessary, the columns of U so that the first n columns are linearly independent, and let W be the $n \times n$ nonsingular matrix satisfying $WU = [I \,|\, U']$. Partition W as $W = [W_1 \,|\, W_2 \,|\, \cdots \,|\, W_k]$, where W_i contains $n_i = \deg P_i$ columns for each $i = 1, 2, \ldots, k$. Define V_{P_i} for each P_i as in Theorem 3.2, a sub-space of G^{n_i} with $\dim V_{P_i} < n_i$. Since W is nonsingular, each W_i has rank n_i, and therefore each W_i has a row that is not in V_{P_i}. Let $w_{i;j}$ denote the jth row of W_i and let s_j be the cardinality of the set $\{i \,|\, w_{i;j} \notin V_{P_i}\}$.

Assume $s_1 = s \geq 1$ and that $w_{i;1} \notin V_{P_i}$, $i = 1, 2, \ldots, s$. Choose $k-s$ rows of W, $j_{s+1}, j_{s+2}, \ldots, j_k$, that satisfy $w_{r;j_r} \notin V_{P_{j_r}}$, $r = s+1, s+2, \ldots, k$. A row vector β with entries in G that are nonzero only in positions $1, j_{s+1}, j_{s+2}, \ldots, j_k$ exists such that if $\gamma = \beta W = [\gamma_1 \,|\, \gamma_2 \,|\, \cdots \,|\, \gamma_k]$, then $\gamma_i \notin V_{P_i}$, $i = 1, 2, \ldots, k$. At least n multiplications are necessary to compute the bilinear form $\beta W A(x)y$ since the jth column of $A_i x_i$ is equal to $C_{P_i}^j x_i$, and any nontrivial linear combination of the columns of $\beta W A(x)$ is

$$\sum_{i=1}^{k} \gamma_i (\sum_{j=0}^{n_i-1} \alpha_{ij} C_{P_i}^j) x_i.$$

Again, as in Theorem 3.2, this combination can be zero only if $\gamma_i \sum_{j=0}^{n_i-1} \alpha_{ij} C_{P_i}^j = 0$ for $i = 1, 2, \ldots, k$, which is impossible unless $\alpha_{ij} = 0$, $\forall i, j$, since $\gamma_i \notin V_{P_i}$ for

$i = 1, 2, \ldots, k$. Therefore $\beta WA(x)$ has n linearly independent columns and by Theorem 2.5 the computation of $\beta WA(x)y$ requires at least n m/d steps. Since β has only $k-s+1$ nonzero entries, $k-s+1$ multiplications are contributed by the partition βI, and at most $t-n$ entries in $\beta U'$ are nonzero, thus $\mu(\beta WA(x)y) \geq k-s+1+t-n$. Therefore $k-s+1+t-n \geq n$, yielding $t \geq 2n-k-1+s$. Since an algorithm has been developed that uses $2n-k$ multiplications to compute $A(x)y$ and $s \geq 1$, we obtain $t = \mu_B(MPM(P); G) = \bar{\mu}_B(MPM(P); G) = 2n-k$. ∎

A common computation to which Theorem 3.3 applies is the evaluation of cyclic (or circular) convolutions. The *cyclic convolution* of two sequences is defined by

$$z_k = \sum_{i=0}^{N-1} x_i y_{k-i}, \quad k = 0, 1, \ldots, N-1, \tag{3.4}$$

where all indexes are reduced to principal residues modulo N. The cyclic convolution of (3.4) is equivalent to the polynomial product

$$z(u) \equiv x(u)y(u) \pmod{u^N-1}$$

where $x(u) = \sum_{i=0}^{N-1} x_i u^i$, $y(u) = \sum_{i=0}^{N-1} y_i u^i$, and $z(u) = \sum_{i=0}^{N-1} z_i u^i$.

The polynomial u^N-1 factors over Q into irreducible cyclotomic polynomials. This factorization is

$$u^N-1 = \prod_{d|N} C_d(u),$$

where $C_d(u)$ is the dth cyclotomic polynomial. The notation $d|N$ means d divides N, and $\prod_{d|N}$ is a shorthand notation for a product over all positive divisors of N. The number of irreducible factors of u^N-1 over Q is therefore equal to the number of positive divisors of N, denoted by $\tau(N)$. Appendix A provides additional information on cyclotomic polynomials.

Corollary 3.1. $\mu_B(MPM(u^N-1); Q) = 2N-\tau(N)$.

Proof. Theorem 3.3 states that $\mu_B(MPM(P); Q) = 2N-k$, where k is the number of distinct irreducible factors of P over G, given that G has sufficient cardinality. The field $G = Q$ is infinite and the number of distinct irreducible factors of u^N-1 over Q is $\tau(N)$, thus

$$\mu_B(MPM(u^N-1); Q) = 2N - \tau(N). \quad ∎$$

3.4. Products of a Fixed Polynomial with Several Polynomials

Sometimes one polynomial is to be multiplied by several other distinct polynomials. The discrete Fourier transform of a sequence whose length is a power of a prime number is equivalent to a union of systems of polynomial multiplication modulo irreducible polynomials, in which some of the blocks repeat. A system of bilinear forms with this structure is

$$
\begin{bmatrix}
A(x_1) & 0 & 0 & 0 \\
0 & A(x_1) & 0 & 0 \\
0 & 0 & B(x_2) & 0 \\
0 & 0 & 0 & B(x_3)
\end{bmatrix}
\begin{bmatrix}
y_1 \\
y_2 \\
y_3 \\
y_4
\end{bmatrix}.
$$

This system is not a direct sum because $A(x_1)$ is shared by two components of the union. Even if the system had only one $A(x_1)$ block, we could not apply Theorem 3.3 to analyze the multiplicative complexity since two of the remaining blocks are products in the same algebra (B is shared). Another possibility not covered by Theorem 3.3 is when two blocks have the same x vector, but involve products in different algebras, as in

$$
\begin{bmatrix}
A(x) & 0 \\
0 & B(x)
\end{bmatrix}
\begin{bmatrix}
y_1 \\
y_2
\end{bmatrix}.
$$

The principal result of this section is that a lower bound on the multiplicative complexity of these systems exists that is equal to the sum of lower bounds on the multiplicative complexity of each of the components. These systems are analyzed as semilinear systems, so the notation $A(f)$ will be used rather than $A(x)$, indicating that entries of the vector f are taken from the field F. One shortcoming of the analysis of this section is that it applies only to irreducible polynomial moduli and excludes powers of irreducible polynomials. This is not a glaring deficiency since no systems encountered in practice include polynomial powers.

3.4.1. Equivalent Systems and Row Reduced Forms

Two systems are considered equivalent if an algorithm for computing one system can be used to compute the other system and vice versa, such that no additional m/d steps are required to convert either system to the other. The following is a rewording of Definition 2.9 that expresses the equivalence of semilinear systems.

Definition 3.1. [6] *Let* $B = F \cup \{y_1, y_2, \ldots, y_m\} \cup \{z_1, z_2, \ldots, z_n\}$. *The two systems* $M(f)y$ *and* $N(f)z$ *are equivalent if two mappings,* $\alpha: \{y_1, y_2, \ldots, y_m\} \to L_G(B)$ *and* $\beta: \{z_1, z_2, \ldots, z_n\} \to L_G(B)$ *exist such that* $L_G(r(M(f)\alpha(y))) = L_G(r(N(f)z))$ *and* $L_G(r(N(f)\beta(z))) = L_G(r(M(f)y))$.

The row rank, $\rho_r(M(f))$, is often less than the actual number of rows of $M(f)$. Such a system can be reduced to an equivalent system with all rows G-linearly independent. A system with all rows G-linearly independent is said to be in *row reduced* form.

3.4.2. Reduction and Inflation Mappings

The concepts presented in §3.2 will now be extended to aid in developing techniques for analyzing unions of systems of polynomial multiplication modulo identical polynomials, and also for unions of such systems in which two or more of the polynomial multiplicands are identical. This analysis will begin with a specification of the algebraic setting of such systems.

Let G be the ground field and u an indeterminate for all fields that occur. The ring $G[u]$ contains all polynomials with coefficients in G. Let K be the quotient ring $K = G[u]/\langle P(u) \rangle$ where $P(u)$ is an irreducible polynomial over G of degree n and $\langle P(u) \rangle$ is the ideal generated by $P(u)$. Elements of K are polynomials in $G[u]$, and each element of K is equivalent to a polynomial in $G[u]$ of degree less than n. Multiplication in K is the normal convolution product used for polynomial multiplication, except that it is common to reduce elements of K to polynomials of degree less than n through reduction modulo $P(u)$. Each nonzero element of K has a multiplicative inverse, thus K is a field and the dimension of K over G is $[K:G] = n$. The usual basis chosen for K is $1, u, u^2, \ldots, u^{n-1}$.

Let $F \supset G$ be a field and let $\mathfrak{R}$ be the ring $\mathfrak{R} = F[u]/\langle P(u) \rangle$. $\mathfrak{R}$ will generally be a field unless $P(u)$ factors over F. The dimension of $\mathfrak{R}$ over F is $[\mathfrak{R}:F] = n$, and $1, u, u^2, \ldots, u^{n-1}$ is a basis of $\mathfrak{R}$. $\mathfrak{R}$ will sometimes be denoted by $\mathfrak{R} = F \otimes K$ to show that it can be constructed as a direct product of algebras.

In more generality $K \supset G$ could be any finite extension field of G such that $[K:G] = n$. Let $u_0, u_1, \ldots, u_{n-1}$ be a basis of K over G, then

$$ku_j = \sum_{i=0}^{n-1} g_{ij} u_i, \quad j = 0, 1, \ldots, n-1,$$

where $g_{ij} \in G$, for $i, j = 0, 1, \ldots, n-1$. Therefore each $k \in K$ can be mapped into an $n \times n$ G-matrix. This mapping, called the *regular representation* of K over G relative to the basis $u_0, u_1, \ldots, u_{n-1}$, will be denoted by $\rho(k) = (g_{ij})$ where (g_{ij}) is a notation for an $n \times n$ G-matrix with entries g_{ij}. When $K = G[u]/\langle P(u) \rangle$, and the basis $1, u, u^2, \ldots, u^{n-1}$ is used, then $\rho(u) = C_P$, where C_P is the companion matrix of $P(u)$, as presented in the proof of Theorem 3.2. The representation ρ can be extended to a representation ρ_F of elements of $\mathfrak{R}$ as $n \times n$ F-matrices.

Let V_K be a K-vector space with $\dim_K V_K = l$. The mapping ρ allows vectors in V_K to be represented as vectors in a G-vector space that will be denoted by V_G. If $v_0, v_1, \ldots, v_{l-1}$ is a basis of V_K and $u_0, u_1, \ldots, u_{n-1}$ is a basis of K over G, then $v_i u_j$, $i = 0, 1, \ldots, l, j = 0, 1, \ldots, n$ is a basis of V_G and $\dim V_G = ln$.

Let V_K and W_K be vector spaces over K with bases $v_0, v_1, \ldots, v_{r-1}$ and $w_0, w_1, \ldots, w_{s-1}$ respectively. The space $\text{Hom}(V_K, W_K)$ contains all linear transformations from V_K to W_K relative to the given bases and is the space of all $s \times r$

matrices over K. Every K-linear transformation $t: V_K \to W_K$ is also a G-linear transformation $V_G \to W_G$. This mapping of an $s \times r$ matrix in K to an $sn \times rn$ matrix in G is denoted by

$$R: \text{Hom}(V_K, W_K) \to \text{Hom}(V_G, W_G)$$

and is called a *reduction mapping* since it reduces the field of definition. If ρ is the regular representation of K relative to the chosen basis of K over G, then the reduction mapping R replaces each entry k_{ij} of an $s \times r$ K-matrix $A = (k_{ij})$ with the $n \times n$ G-matrix $\rho(k_{ij})$ in forming $R(A)$.

An inverse mapping, called the *inflation mapping*, exists, but only applies when each $n \times n$ submatrix of the $sn \times rn$ G-matrix is equal to $\rho(k_{ij})$ for some $k_{ij} \in K$. The inflation mapping is

$$\Im: (g_{\alpha\beta}) \to (k_{ij}).$$

Each element of K^t can be viewed as a vector in G^{nt}, yielding the isomorphism

$$\Im_s: G^{nt} \to K^t.$$

Let $B_G \in R(\text{Hom}(V_K, W_K))$, and let $a \in G^{nr}$, then

$$\Im_s(B_G a) = \Im(B_G) \Im_s(a).$$

$\Im_s$ is called the *pseudo-inflation mapping*. This mapping can be extended to a mapping from $tn \times l$ G-matrices to $t \times l$ K-matrices by applying the mapping to each of the columns of the G-matrix.

A *pseudo-reduction mapping* $R_s: V_K \to V_G$ can be defined that is the inverse of $\Im_s$ and can be similarly extended to a mapping from $t \times l$ K-matrices to $nt \times l$ G-matrices.

The concepts of reduction and inflation mappings can be applied to the ring $\Re = F \otimes K$. The reduction mapping replaces each entry r_{ij} of an $s \times r$ $\Re$-matrix with $\rho_F(r_{ij})$, yielding an $sn \times rn$ F-matrix. The inflation mapping from $sn \times rn$ F-matrices to $s \times r$ $\Re$-matrices exists under conditions similar to those specified for the inflation mapping from G-matrices to K-matrices.

3.4.3. Equivalence of Products with a Fixed Polynomial

The previous two sections have provided an algebraic framework and valuable tools for the analysis of systems consisting of a union of products of a single polynomial by several polynomials.

Let $F \supset G$ be a field and let $H = F(y_1, y_2, \ldots, y_m)$, where $y_1, y_2, \ldots, y_m$ are indeterminates. This system description is identical to those proposed in Chapter 2, where F is typically $F = G(x_1, x_2, \ldots, x_r)$ for some indeterminates $x_1, x_2, \ldots, x_r$. As in the previous section, let $K \supset G$ be a finite extension of G with $u_0, u_1, \ldots, u_{n-1}$ a basis of K over G. Let $\Re = F \otimes K$ and $\Re(y) = H \otimes K$. Each $r \in \Re$ can be expressed as $\sum_{i=0}^{n-1} f_i u_i$, where $f_i \in F$, $i = 0, 1, \ldots, n-1$. Similarly, each element of $\Re(y)$ can be expressed as $\sum_{i=0}^{n-1} h_i u_i$, where $h_i \in H$, $i = 0, 1, \ldots, n-1$.

Consider the system

$$(\sum_{i=0}^{n-1} f_i u_i)(\sum_{i=0}^{n-1} l_{\alpha i} u_i) = \sum_{i=0}^{n-1} \xi_{\alpha i} u_i; \quad \alpha = 1, 2, \ldots, t,$$

where $f_i \in F$, $l_{\alpha i} \in L_G(y) \subseteq \Re(y)$. Auslander and Winograd [6] have chosen to denote the system $\{\xi_{\alpha 0}, \xi_{\alpha 1}, \ldots, \xi_{\alpha(n-1)}\}$ by $C(K; f, l_\alpha; \{u_i\})$, where $f = \sum_{i=0}^{n-1} f_i u_i \in \Re$ and $l_\alpha = \sum_{i=0}^{n-1} l_{\alpha i} u_i \in L_G(y) \otimes K \subseteq \Re(y)$. When $K = G[u]/\langle P(u) \rangle$, where $P(u)$ is an irreducible polynomial of degree n, then $C(K; f, l_\alpha; \{u_i\})$ may be denoted by $C(P; f, l_\alpha)$ where the basis $u_i = u^i$ is assumed. We will begin by examining the system $\bigcup_{\alpha=1}^{t} C(K; f, l_\alpha; \{u_i\})$.

Let ρ be the regular representation of K over G relative to the basis $u_0, u_1, \ldots, u_{n-1}$. Let $A(f)$ be the $n \times n$ F-matrix $A(f) = \rho_F(f) = \sum_{i=0}^{n-1} f_i \rho(u_i)$. Let $l_\alpha = [l_{\alpha 0} \, l_{\alpha 1} \cdots l_{\alpha(n-1)}]^T$ and $\xi_\alpha = [\xi_{\alpha 0} \, \xi_{\alpha 1} \cdots \xi_{\alpha(n-1)}]^T$. Then $\xi_\alpha = A(f) l_\alpha$ and the matrix representation of the system under consideration is

$$I = \begin{bmatrix} \xi_1 \\ \xi_2 \\ \vdots \\ \xi_t \end{bmatrix} = \begin{bmatrix} A(f) & 0 & \cdots & 0 \\ 0 & A(f) & \cdots & 0 \\ \vdots & \vdots & \ddots & \vdots \\ 0 & 0 & \cdots & A(f) \end{bmatrix} \begin{bmatrix} l_1 \\ l_2 \\ \vdots \\ l_t \end{bmatrix}.$$

When $K = G[u]/\langle P(u) \rangle$ and the basis $1, u, \ldots, u^{n-1}$ is used, then $\rho(u) = C_P$, the companion matrix of $P(u)$, and $A(f) = \sum_{i=0}^{n-1} f_i C_P^i$.

Since $l_{\alpha i} \in L_G(y)$, then $l_{\alpha i} = \sum_{s=1}^{m} g_{\alpha i s} y_s$, $g_{\alpha i s} \in G$, and

$$l_\alpha = (g_{\alpha is}) \begin{bmatrix} y_1 \\ y_2 \\ \vdots \\ y_m \end{bmatrix} = M_\alpha y,$$

where M_α is an $n \times m$ G-matrix. Let $M = [M_1 \ M_2 \ \cdots \ M_t]^T$ be an $nt \times m$ G-matrix, then

$$I = \begin{bmatrix} \xi_1 \\ \xi_2 \\ \vdots \\ \xi_t \end{bmatrix} = \begin{bmatrix} A(f) & 0 & \cdots & 0 \\ 0 & A(f) & \cdots & 0 \\ \vdots & \vdots & \ddots & \vdots \\ 0 & 0 & \cdots & A(f) \end{bmatrix} My.$$

The system $I = \overset{t}{\underset{\alpha=1}{\cup}} C(K;f,l_\alpha;\{u_i\})$ will be denoted by $I = (tA(f))My$ to emphasize that each of the $A(f)$ blocks are identical.

If another basis $v_0, v_1, \ldots, v_{n-1}$ for K over G were selected, then $I' = \overset{t}{\underset{\alpha=1}{\cup}} C(K;f,l_\alpha;\{v_i\}) = (tA'(f))M'y$. Let $l{:}K \to K$ be the G-linear mapping that changes basis from v_i to u_i, defined by $l(v_i) = u_i, i = 0, 1, \ldots, n-1$. If L is the matrix representing this mapping, then $A'(f) = L^{-1}A(f)L$ and $M'_\alpha = L^{-1}M_\alpha$. Consequently

$$I' = \begin{bmatrix} A'(f) & 0 & \cdots & 0 \\ 0 & A'(f) & \cdots & 0 \\ \vdots & \vdots & \ddots & \vdots \\ 0 & 0 & \cdots & A'(f) \end{bmatrix} M'y$$

(3.5)

$$= \begin{bmatrix} L^{-1} & 0 & \cdots & 0 \\ 0 & L^{-1} & \cdots & 0 \\ \vdots & \vdots & \ddots & \vdots \\ 0 & 0 & \cdots & L^{-1} \end{bmatrix} \begin{bmatrix} A(f) & 0 & \cdots & 0 \\ 0 & A(f) & \cdots & 0 \\ \vdots & \vdots & \ddots & \vdots \\ 0 & 0 & \cdots & A(f) \end{bmatrix} My$$

and therefore I and I' are equivalent systems. Hence the basis of K over G is irrelevant in evaluating the multiplicative complexity of these systems, and $C(K;f,l_\alpha;\{u_i\})$ will now be denoted by $C(K;f,l_\alpha)$ to reflect this equivalence of identical systems with different bases.

Let $K' \supset K$ be a finite extension of K with $v_0, v_1, \ldots, v_{p-1}$ a basis of K' over K and assume $v_0 = 1$. The injection mapping $i : K \rightarrow K'$ allows elements of K to be represented in K'. This mapping can be extended to represent elements of $\Re(y)$ as elements of $\Re'(y) = H \otimes K'$ where each element $\sum\limits_{i=0}^{n-1} h_i u_i$ of $\Re(y)$ is mapped into the element $\sum\limits_{i=0}^{n-1} h_i \cdot (v_0 u_i)$ of $\Re'(y)$.

Define the system I'' by

$$I'' = \bigcup_{\alpha=1}^{t} C(K'; f, l_\alpha)$$

where f and l_α are now the images of the original f and l_α in $\Re'(y)$. Let ρ be the regular representation of K over G relative to the basis $u_0, u_1, \ldots, u_{n-1}$ and ρ' be the regular representation of K' over K relative to the basis $v_0, v_1, \ldots, v_{p-1}$. Let ρ'' be the regular representation of K' over G relative to the lexicographically ordered set $\{v_i u_j\}$. If $f \in \Re'(y)$ is $f = \sum\limits_{i=0}^{n-1} f_i (v_0 u_i) = (\sum\limits_{i=0}^{n-1} f_i u_i) v_0$, then since $v_0 = 1$, $\rho'(f) = (\sum\limits_{i=0}^{n-1} f_i u_i) I_p$, where I_p is the $p \times p$ identity matrix. $\rho''(f)$ is obtained by reducing the field of definition, therefore

$$\rho''(f) = \begin{bmatrix} A(f) & 0 & \cdots & 0 \\ 0 & A(f) & \cdots & 0 \\ \vdots & \vdots & \ddots & \vdots \\ 0 & 0 & \cdots & A(f) \end{bmatrix}$$

and $C(K'; f, l_\alpha)$ is

$$\begin{bmatrix} A(f) & 0 & \cdots & 0 \\ 0 & A(f) & \cdots & 0 \\ \vdots & \vdots & \ddots & \vdots \\ 0 & 0 & \cdots & A(f) \end{bmatrix} \begin{bmatrix} M_\alpha \\ 0 \\ \vdots \\ 0 \end{bmatrix} y = \begin{bmatrix} I \\ 0 \\ \vdots \\ 0 \end{bmatrix} A(f) M_\alpha y. \tag{3.6}$$

Therefore $C(K'; f, l_\alpha)$ is equivalent to $C(K; f, l_\alpha)$ and thus the systems I and I'' are also equivalent.

The multiplicative complexity of $C(K; f, l_\alpha)$ does not depend on whether f or l_α are viewed as elements of $\Re(y)$ or $\Re'(y)$. Since the system I computes the coordinates of $f \cdot l_\alpha$, then the system $C(K; f, l_\alpha)$ will be referred to as $C(f \cdot l_\alpha)$ unless the field K is to be emphasized.

The following example will clarify some of the concepts presented in the past few sections.

Example 3.3. Let $G = Q$, the field of rational numbers, $K = G[u]/\langle u^2+1 \rangle$ with natural basis $1, u$, and $F = G(\sqrt{2})$. Consider the system $C(f \cdot l)$, where $f = fk$, $f = \sqrt{2} \in F$, $k = 1-u \in K$. Thus $f = \begin{bmatrix} \sqrt{2} \\ -\sqrt{2} \end{bmatrix}$ is a representation of f as a vector in F^2. Let y have dimension 1 and let the single entry of y be $y_0 = \sqrt{3}$. Let the 2×1 G-matrix M be

$$M = \begin{bmatrix} 1 \\ 2 \end{bmatrix}.$$

Elements of the field K can be represented as an ordered pair of elements of G, where multiplication of two pairs follows the rules of complex multiplication. The regular representation is $\rho(u) = C_P = \begin{bmatrix} 0 & -1 \\ 1 & 0 \end{bmatrix}$, and thus $A(f) = [f \ \ C_P f]$. Clearly $t = 1$, and the complete system is

$$A(f)My = \begin{bmatrix} \sqrt{2} & \sqrt{2} \\ -\sqrt{2} & \sqrt{2} \end{bmatrix} \begin{bmatrix} 1 \\ 2 \end{bmatrix} [\sqrt{3}] = \begin{bmatrix} 3\sqrt{2} \\ \sqrt{2} \end{bmatrix} \sqrt{3}.$$

This system is not row reduced since the two rows of $A(f)M$ are scalar multiples of each other in G. For this system $\rho_r(A(f)M) = \rho_c(A(f)M) = 1$, thus $\mu(A(f)My) \geq 1$ and since one multiplication, $\sqrt{2} \cdot \sqrt{3}$, suffices to compute the system then $\mu(A(f)My) = 1$.

The ability to reduce the system $C(K'; f, l_\alpha)$ to an equivalent system over K is one example of reducing a system to a minimal equivalent system, in the sense that t is minimal. Given two systems, $\overset{t}{\underset{\alpha=1}{\cup}} C(f \cdot l_\alpha)$ and $\overset{s}{\underset{\beta=1}{\cup}} C(f \cdot l'_\beta)$, it can be shown that they are equivalent if $L_G(M_\alpha) = L_G(M'_\beta)$. Thus a system $\overset{t}{\underset{\alpha=1}{\cup}} C(f \cdot l_\alpha)$ may be reduced to an equivalent system with the minimum value of t, that is, $t = \dim L_G(M_\alpha)$. From the previous example we see that a system may satisfy this condition yet not be row reduced. A more general result will now be proven regarding the equivalence of systems of the type being considered.

Theorem 3.4. *Let $(tA(f))My$ and $(sA(f))M'y$ be two systems. Let $L_K^r(M)$ be the K-linear span of the rows of $\mathfrak{I}_s(M)$ and $L_K^r(M')$ be the K-linear span of $\mathfrak{I}_s(M')$. If $L_K^r(M) = L_K^r(M')$ then $(tA(f))My$ and $(sA(f))M'y$ are equivalent.*

Proof. Let B be the $s \times t$ K-matrix satisfying $\mathfrak{I}_s(M') = B\mathfrak{I}_s(M)$ and let $B_G = R(B)$ be the $sn \times tn$ G-matrix obtained by reducing the field of definition. Clearly $B_G M = M'$ and therefore

$$B_G(tA(f))My = (sA(f))B_GMy = (sA(f))M'y$$

and if $\Im_s(M) = B'\Im_s(M')$ then

$$B'_G(sA(f))M'y = (tA(f))My. \quad \blacksquare$$

In the proof of this theorem the assumption was made that F is a commutative field and thus that $\Re$ is a commutative ring. If not, then the identity $B_G(tA(f)) = (sA(f))B_G$ would not necessarily be true. The identity can be easily verified in the ring F since $\Im((tA(f))) = rI_t$ and $\Im((sA(f))) = rI_s$ where I_t and I_s are the $t \times t$ and $s \times s$ identity matrices in F, respectively, and $r \in \Re$ is f as an element of $\Re$. Therefore $BrI_t = rI_sB$, or $Br = rB$, implying the commutativity of multiplication in F and $\Re$. Application of the inflation mapping $\Im$ gives the desired identity. The following definition formalizes this concept of row reduction over K.

Definition 3.2. [6] *A system* $\overset{t}{\underset{\alpha=1}{\cup}} C(f \cdot l_\alpha)$ *is said to be quasi-row reduced if the t rows of* $\Im_s(M)$ *are linearly independent over K.*

The concept of column reduction can also be applied here. The system $(tA(f))My$ is column reduced if all the columns of M are G-linearly independent. Since $\Im_s(MC) = \Im_s(M)C$ for any $m \times l$ G-matrix C, then $(tA(f))My$ is column reduced if all the columns of $\Im_s(M)$ are G-linearly independent.

Definition 3.3. [6] *A system* $\overset{t}{\underset{\alpha=1}{\cup}} C(f \cdot l_\alpha)$ *is said to be quasi-row-column reduced (qrc reduced) if the t rows of* $\Im_s(M)$ *are linearly independent over K and the m columns of* $\Im_s(M)$ *are linearly independent over G.*

A system $(tA(f))My$ can always be converted to an equivalent system $(t'A(f))M'y'$ that is qrc reduced. The system $(t'A(f))M'y'$ can be obtained by selecting a $t' \times m'$ submatrix of $\Im_s(M)$ whose rows are K-linearly independent and whose columns are G-linearly independent, where $t' = \dim L_K^r(M)$ and m' is the dimension of the G-linear span of the columns of the quasi-row reduced $\Im_s(M)$. The resulting $t' \times m'$ matrix is $\Im_s(M')$ and thus $M' = R_s(\Im_s(M'))$.

In this section it has been shown that a semilinear system of the form $(tA(f))My$ is equivalent to the qrc reduced system $(sA(f))M'y$ where $M' = B_GMC$, $B_G = R(B)$ is a $tn \times tn$ G-matrix, B is a $t \times t$ K-matrix, and C is an $m \times m$ G-matrix. A simple example will show this concept.

Example 3.4. Let $G = Q$, $K = G[u]/\langle u^2+1\rangle$, and $F = G(x_1, x_2)$ where x_1 and x_2 are indeterminates. Let y_1, y_2, y_3 be three indeterminates. Define the system S as

$$S = \begin{bmatrix} \xi_1 \\ \xi_2 \end{bmatrix} = A(f)My = \begin{bmatrix} x_1 & -x_2 \\ x_2 & x_1 \end{bmatrix} \begin{bmatrix} 1 & 2 & 3 \\ 0 & 1 & 2 \end{bmatrix} \begin{bmatrix} y_1 \\ y_2 \\ y_3 \end{bmatrix}.$$

For this example $t = 1$ and M is obviously not qrc reduced.

M is row reduced since $L_K^r(M) = t = 1$, but is not quasi-column reduced since there are more columns than rows. The column rank of M is 2, thus S is equivalent to a system $A(f)M'y'$ where M' is 2×2 and y' has only two distinct indeterminates. It is instructive to construct the equivalent system by finding an invertible G-matrix C, such that MC has a zero vector as a column. Suppose that it is convenient to have $M' = I_2$, the 2×2 identity matrix, then

$$C = \begin{bmatrix} 1 & -2 & 1 \\ 0 & 1 & -2 \\ 0 & 0 & 1 \end{bmatrix}$$

yields $MC = [M' \,|\, 0]$.

This system will compute S by simply redefining the indeterminates in y as $y' = NC^{-1}y$, where $N = [I_2 \,|\, 0]$. Inverting C yields

$$C^{-1} = \begin{bmatrix} 1 & 2 & 3 \\ 0 & 1 & 2 \\ 0 & 0 & 1 \end{bmatrix}$$

and

$$y' = \begin{bmatrix} y_1' \\ y_2' \end{bmatrix} = \begin{bmatrix} y_1 + 2y_2 + 3y_3 \\ y_2 + 2y_3 \end{bmatrix}.$$

Thus

$$S = \begin{bmatrix} \xi_1 \\ \xi_2 \end{bmatrix} = A(f)M'y' = \begin{bmatrix} x_1 & -x_2 \\ x_2 & x_1 \end{bmatrix} \begin{bmatrix} 1 & 0 \\ 0 & 1 \end{bmatrix} \begin{bmatrix} y_1' \\ y_2' \end{bmatrix}.$$

is a representation of the original system with one less indeterminate, and has the same multiplicative complexity as the original system.

3.4.4. Multiplicative Complexity Results

Based on the previous several sections, any system of the form $(tA(f))My$ can be reduced to an equivalent system with minimal row rank over K and minimal column rank over G. In the following analysis it will be assumed that this reduction has been performed and that any system being considered is therefore qrc reduced.

The number of m/d steps necessary to compute $(tA(f))My$ can be determined by induction on t and m, where m is the number of distinct indeterminates in y. We will show that for $m > t$, the multiplicative complexity of the system $(tA(f))My$ is related to the multiplicative complexity of systems of the form $((t-1)A(f))M'y'$ or $(tA(f))M'y'$, where $y' = [y_2\, y_3\, \cdots\, y_m]^T$. A system $(tA(f))My$ can be computed using an algorithm for one of these two systems and some additional m/d steps, whose number is bounded from below. The following lemma provides a means of identifying which of the two types of systems are obtained in reducing a given $(tA(f))My$.

Lemma 3.2. [6] *Let A be a $t \times s$ K-matrix of rank t and let B be an $s \times (s-1)$ G-matrix of rank $s-1$. If $v \in G^s$ satisfies $vB = 0$, then $\operatorname{rank} AB = t-1$ if and only if $K \otimes v$ is in the K-linear span of the rows of A, otherwise $\operatorname{rank} AB = t$.*

Proof. Since $\operatorname{rank} B = s-1$, then either $\operatorname{rank} AB = t$ or $\operatorname{rank} AB = t-1$. Assume $\operatorname{rank} AB = t-1$, then a nonzero row vector $v_K \in K^t$ exists such that $v_K AB = 0$. Let $w_K = v_K A \neq 0$, since $\operatorname{rank} A = t$. As a K-matrix the rank of B is also $s-1$, therefore $\{u_K \in K^s \mid u_K B = 0\}$ is a one-dimensional subspace of K^s that must be $K \otimes v$ since $vB = 0$. $w_K B = 0$ implies $w_K \in K \otimes v$. But $w_K = v_K A$, so w_K is in the K-linear span of the rows of A, and therefore $K \otimes v$ is in the K-linear span of the rows of A.

To prove the other half of the relation, assume $K \otimes v$ is in the K-linear span of the rows of A. Let C be a $t \times t$ invertible K-matrix such that the first row of CA is v. The first row of CAB will be 0 and therefore the rank of CAB is less than t. Since C is nonsingular the rank of CAB is equal to the rank of $C^{-1}CAB = AB$ and must be $t-1$. ∎

Let G be an infinite field, $F \supset G$ an extension field of G, and $K \supset G$ a finite extension of G, unrelated to F. Let $u_0, u_1, \ldots, u_{n-1}$ be a basis of K over G and let $f = \sum_{i=0}^{n-1} f_i u_i \in \Re$. Let $s = \dim L_G(r(f_0), r(f_1), \ldots, r(f_{n-1}))$ where $r : F \to F/G$ is the natural homomorphism and s will be referred to as the *row rank of f*. Let $y_1, y_2, \ldots, y_m$ be a set of distinct indeterminates comprising the entries of y. The *cardinality of y* is said to be m, indicating that y consists of m distinct indeterminates.

Theorem 3.5. [6] *If $(tA(f))My$ is a qrc reduced system with the cardinality of y equal to m and the row rank of f equal to s, $s \geq 1$, then*

$$\mu_B((tA(f))My; G) \geq t(s-1)+m$$

where $B = F \cup \{y_1, y_2, \ldots, y_m\}$.

Proof. Consider first the case $t = 1$, $m = 1$, which is the system $A(f)My$. For this system $A(f) = \sum_{i=0}^{n-1} f_i C_i$, where $C_i = \rho(u_i)$ is the regular representation of u_i, M is an $n \times 1$ G-matrix, and y is an indeterminate. Let $k = \Im_s(M)$ and let $B = \rho(k^{-1})$. $A(f)My$ is equivalent to $BA(f)My = A(f)BMy$. We can assume without loss of generality that $u_0 = 1$, in which case $BM = [1 \; 0 \; \cdots \; 0]^T$ and $A(f)BM = [f_0 \; f_1 \; \cdots \; f_{n-1}]^T$. Thus $\rho_r(A(f)BM) = s$ and Theorem 2.4 yields

$$\mu_B(A(f)My) \geq s = 1 \cdot (s-1) + 1.$$

The theorem has now been proven for $(t, m) = (1, 1)$ and now we will assume that it is true for all $(t', m') < (t, m)$ where the set of ordered pairs (t, m) is ordered lexicographically. We will show that for any (t, m) the system $(tA(f))My$ can be modified to yield a smaller system of the same form requiring fewer m/d steps than $(tA(f))My$. The difference between the number of m/d steps required for this smaller system and the original system can be bounded and the lower bound is always greater than zero, hence this type of reduction of a system can be recursively applied to obtain a lower bound on the number of m/d steps required to compute $(tA(f))My$.

Two types of reduction will be performed on $(tA(f))My$. The first reduction is used when a nonsingular K-matrix T exists such that $T\Im_s(M)$ has a row of the form $k[g_1 \, g_2 \; \cdots \; g_m], g_i \in G, k \in K$.

The system $(tA(f))My$ may be replaced by the equivalent system $R(T)(tA(f))My = (tA(f))M'y$ where $\Im_s(M') = T\Im_s(M)$. Assume, without loss of generality, that the first row of $\Im_s(M')$ is $k[g_1 \, g_2 \; \cdots \; g_m] = k \otimes v, k \neq 0, v \in G^m$. Let $D = \mathrm{diag}(k^{-1}, 1, 1, \ldots, 1)$ be a $t \times t$ K-matrix, and let E be a nonsingular $m \times m$ G-matrix satisfying $vE = [1 \; 0 \; \cdots \; 0]$. If $M'' = R_s(\Im_s(M''))$, where $\Im_s(M'') = D\Im_s(M')E$, then the systems $(tA(f))M''y$ and $(tA(f))My$ are equivalent. Since the first row of $\Im_s(M'')$ is $[1 \; 0 \; \cdots \; 0]$, then the first n outputs of $(tA(f))M''y$ are $f_i y_1$, $i = 0, 1, \ldots, n-1$. This is exactly the structure of the system analyzed in Theorem 2.6, from which we conclude that $\mu_B((tA(f))My) = \mu_B(tA(f))M''y \geq \mu_B(((t-1)A(f))\tilde{M}y') + s$ where $\Im_s(\tilde{M})$ is the submatrix of $\Im_s(M'')$ resulting from the omission of the first row and first column, and $y' = [y_2 \, y_3 \; \cdots \; y_m]^T$. By assumption, $((t-1)A(f))\tilde{M}y'$ is qrc reduced and thus must satisfy the induction hypothesis, hence

$$\mu_B((tA(f))My) \geq (t-1)(s-1) + m - 1 + s = t(s-1) + m,$$

which agrees with the hypothesis.

The only remaining possibility is that no nonsingular K-matrix T exists such that $T\Im_s(M)$ has a row of the form $k[g_1 \, g_2 \; \cdots \; g_m], g_i \in G, k \in K$. For this case Theorem 2.5 guarantees the existence of the specialization

$$\alpha(y_1) = \sum_{i=2}^{m} g_i y_i,$$

$$\alpha(y_i) = y_i, \quad i = 2, 3, \ldots, m.$$

Let E be the $m \times (m-1)$ G-matrix whose first row is $[g_1 \, g_2 \, \cdots \, g_m]^T$ and whose last $m-1$ rows are the $(m-1) \times (m-1)$ identity matrix. Clearly $\alpha^*((tA(f))My) = (tA(f))MEy'$, where $y' = [y_2 \, y_3 \, \cdots \, y_m]^T$, and therefore by Theorem 2.5, $\mu_B((tA(f))My) \geq \mu_B((tA(f))MEy') + 1$. Since for this case no matrix T exists satisfying the assumptions, there can be no vector $v \in G^m$ such that $K \otimes v$ is in the K-linear span of the rows of M, by Lemma 3.2 the rank of $\Im_s(ME)$ is t. The columns of $\Im_s(ME)$ are G-linearly independent and thus the system $(tA(f))MEy'$ is qrc reduced. Therefore, since $(tA(f))MEy'$ must satisfy the induction hypothesis for $(t, m-1)$,

$$\mu_B((tA(f))My) \geq t(s-1) + m - 1 + 1 = t(s-1) + m. \quad \blacksquare$$

This theorem is powerful and can be used to provide many of the lower bounds derived earlier in this chapter. Before demonstrating the applications of this theorem it will be useful to generalize first to the case where more than one element $f \in F$ is included in the union, and then to the case where more than one modulus polynomial is used.

3.5. Products with Several Fixed Polynomials in the Same Ring

One question that has not been addressed in the previous sections of this chapter is the multiplicative complexity of several unrelated polynomial products modulo a single irreducible polynomial. The difference between this type of system and that discussed in the previous section is that there are now $f_1, f_2, \ldots, f_J \in F$ rather than a single $f \in F$. A system of this type will be denoted by

$$\bigcup_{j=1}^{J} (t_j A(f_j)) M^{(j)} y.$$

Just as for $(tA(f))My$, the notion of a qrc reduced system will be useful. The analogue of Theorem 3.4 can be proven for this type of system, resulting in the following theorem.

Theorem 3.6. *The two systems* $\bigcup_{j=1}^{J} (t_j A(f_j)) M^{(j)} y$ *and* $\bigcup_{j=1}^{J} (s_j A(f_j)) N^{(j)} z$ *are equivalent if and only if the K-linear span of the rows of $\Im_s(M^{(j)})$ and $\Im_s(N^{(j)})$ are identical for $j = 1, 2, \ldots, J$.*

Proof. The proof of this theorem follows exactly the same reasoning as Theorem 3.4 and will thus be omitted.

Theorem 3.7. *The two systems* $\bigcup_{j=1}^{J} (t_j A(f_j)) M^{(j)} y$ *and* $\bigcup_{j=1}^{J} (t_j A(f_j)) N^{(j)} z$ *are equivalent if and only if two G-matrices S and T exist such that for each $j = 1, 2, \ldots, J$, $M^{(j)} = N^{(j)} S$ and $N^{(j)} = M^{(j)} T$.*

Proof. The proof of this theorem follows the same line of reasoning used in the proof of Theorem 3.4 and will therefore be omitted.

Definition 3.4. *A system* $\bigcup_{j=1}^{J} (t_j A(f_j)) M^{(j)} y$ *is said to be quasi-row-column reduced if for each $j = 1, 2, \ldots, J$ the rows of $\Im_s(M^{(j)})$ are K-linearly independent and if all the columns of*

$$\tilde{M} = \begin{bmatrix} M^{(1)} \\ \vdots \\ M^{(J)} \end{bmatrix}$$

are G-linearly independent.

Based on this definition and the two theorems preceding it, two qrc reduced systems $\bigcup_{j=1}^{J} (t_j A(f_j)) M^{(j)} y$ and $\bigcup_{j=1}^{J} (t_j A(f_j)) N^{(j)} z$ are equivalent if a nonsingular $m \times m$ G-matrix S exists and if for each $j = 1, 2, \ldots, J$ a nonsingular $t_j \times t_j$ K-matrix D_j exists such that $\Im_s(N^{(j)}) = D_j \Im_s(M^{(j)}) S$, $j = 1, 2, \ldots, J$. We are now prepared to determine a lower bound on the multiplicative complexity of $\bigcup_{j=1}^{J} (t_j A(f_j)) M^{(j)} y$.

As for the system $(tA(f)) My$, let G be an infinite field, $F \supset G$ an extension field of G, and $K \supset G$ a finite extension of G, unrelated to F, with basis $u_0, u_1, \ldots, u_{n-1}$ over G. Let $s_j = \dim L_G(r(f_{j_0}), r(f_{j_1}), \ldots, r(f_{j_{n-1}}))$ be the row rank of $f_j = \sum_{i=0}^{n-1} f_{j_i} u_i \in \Re$, $j = 1, 2, \ldots, J$, where $r : F \to F/G$ is the natural homomorphism.

Theorem 3.8. [6] *If $\bigcup_{j=1}^{J} (t_j A(f_j)) M^{(j)} y$ is a qrc reduced system with the cardinality of y equal to m and the row rank of f_j equal to $s_j \geq 1$, $j = 1, 2, \ldots, J$, and if for every $L \subseteq \{1, 2, \ldots, J\}$, $\dim L_G \bigcup_{l \in L} r(f_l) \geq \sum_{l \in L} (s_l - 1) + 1$, then*

$$\mu_B(\bigcup_{j=1}^{J} (t_j A(f_j)) M^{(j)} y; G) \geq \sum_{j=1}^{J} t_j(s_j - 1) + m,$$

where $r(f_l)$ denotes the set $\{r(f_{l_0}), r(f_{l_1}), \ldots, r(f_{l_{n-1}})\}$ and $B = F \cup \{y_1, y_2, \ldots, y_m\}$.

Proof. Let $t = \sum_{j=1}^{J} t_j$ be the total number of blocks in the union. As for Theorem 3.5, the proof will be by induction on the set of pairs (t, m) ordered lexicographically. When $t = 1$, then J must be 1, and the result of Theorem 3.5 provides a starting point for the induction.

As in the proof of Theorem 3.5, assume the theorem true for all $(t', m') < (t, m)$. Two reductions will be used that are analogous to those used in the proof of Theorem 3.5. The first reduction is used when a nonsingular K-matrix T_j exists such that $T_j \mathfrak{S}_s(M^{(J)})$ has a row of the form $k[g_1 g_2 \cdots g_m]$, $g_i \in G$, $k \in K$ for some $j \in \{1, 2, \ldots, J\}$.

For this case, as for Theorem 3.5, assume without loss of generality that the first row of $T_j \mathfrak{S}_s(M^{(J)})$ is $k[g_1 g_2 \cdots g_m] = k \otimes v \in K \otimes G^m$. Define L as the subset of $\{1, 2, \ldots, J\}$ consisting of all l for which there exists T_l such that the first row of $T_l \mathfrak{S}_s(M^{(l)})$ is $k_l \otimes v$, where $k_l \in K$ for $l \in L$. The original system $\underset{j=1}{\overset{J}{\cup}} (t_j A(f_j)) M^{(J)} y$ is equivalent to the system $\underset{j=1}{\overset{J}{\cup}} (t_j A(f_j)) M'^{(J)} y$, where $\mathfrak{S}_s(M'^{(J)}) = \mathfrak{S}_s(M^{(J)})$ for $j \notin L$ and $\mathfrak{S}_s(M'^{(J)}) = T_j \mathfrak{S}_s(M^{(J)})$ for $j \in L$. Let D be the diagonal $t \times t$ K-matrix whose diagonal entries are $D_{jj} = 1$, $j \notin L$, and $D_{jj} = k^{-1}$, $j \in L$. Let E be a nonsingular $m \times m$ G-matrix satisfying $vE = [1 \ 0 \ \cdots \ 0]$ and let $\overline{M}^{(J)} = D_{jj} M'^{(J)} E$, $j = 1, 2, \ldots, J$. The original system $\underset{j=1}{\overset{J}{\cup}} (t_j A(f_j)) M^{(J)} y$ is equivalent to $\underset{j=1}{\overset{J}{\cup}} (t_j A(f_j)) \overline{M}^{(J)} y$.

Since the first row of $\mathfrak{S}_s(\overline{M}^{(J)})$ is $[1 \ 0 \ \cdots \ 0]$ for $j \in L$, each output of these blocks is of the form $f_j y_1$. Therefore by Theorem 2.6,

$$\mu_B(\underset{j=1}{\overset{J}{\cup}} (t_j A(f_j)) M^{(J)} y) \geq \mu_B \left[\alpha^* \left[\underset{j=1}{\overset{J}{\cup}} (t_j A(f_j)) \overline{M}^{(J)} y \right] \right] + d$$

where $d = \dim L_G(\underset{l \in L}{\cup} r(f_l))$ and α is the specialization of y_1 defined by

$$\alpha(y_1) = 0$$

$$\alpha(y_i) = y_i, \quad i = 2, 3, \ldots, m.$$

Each subsystem $\alpha^*((t_j A(f_j)) \overline{M}^{(J)} y)$ is identical to $(t_j A(f_j)) N^{(J)} y'$ where $y' = [y_2 y_3 \cdots y_m]^T$, $N^{(J)} = \overline{M}^{(J)} B$, and

$$B = \left[\begin{array}{c} 0 \ 0 \ \cdots \ 0 \\ \hline I \end{array} \right]$$

is an $m \times (m-1)$ G-matrix.

Lemma 3.2 states that the rank of $\mathfrak{S}_s(N^{(J)})$ is $(t_j - 1)$ for every $j \in L$ and is t_j for every $j \notin L$. Let $\overline{N}^{(J)}$ be such that $\mathfrak{S}_s(\overline{N}^{(J)})$ is $\mathfrak{S}_s(N^{(J)})$ without the first row for every

$j \notin L$ and $\Im_s(\overline{N}^{(j)})$ is $\Im_s(N^{(j)})$ for every $j \in L$. Also let $t'_j = t_j - 1$ for $j \in L$ and $t'_j = t_j$ for $j \notin L$.

From Lemma 3.2 we know that $\underset{j=1}{\overset{J}{\cup}}(t'_j A(f_j))\overline{N}^{(j)}y'$ is qrc reduced and is equivalent to $\alpha^* \left[\underset{j=1}{\overset{J}{\cup}}(t_j A(f_j))\overline{M}^{(j)}y \right]$. The total number of blocks in the union for this system is $t' = t - \underset{l \in L}{\sum} 1$ and the cardinality of y is $m' = m - 1$, thus $(t', m') < (t, m)$ and the induction hypothesis applies. Therefore

$$\mu_B \left[\underset{j=1}{\overset{J}{\cup}}(t'_j A(f_j))\overline{N}^{(j)}y' \right] \geq \left[\sum_{j=1}^{J} t'_j(s_j - 1) \right] + m - 1$$

$$= \sum_{j=1}^{J} t_j(s_j - 1) + m - 1 - \sum_{l \in L}(s_l - 1).$$

In the statement of the theorem it was assumed that

$$d = \dim L_G \left[\underset{l \in L}{\cup} r(f_l) \right] \geq \sum_{l \in L}(s_l - 1) + 1,$$

yielding

$$\mu_B \left[\underset{j=1}{\overset{J}{\cup}}(t_j A(f_j))M^{(j)}y \right] \geq \mu_B \left[\underset{j=1}{\overset{J}{\cup}}(t'_j A(f_j))\overline{N}^{(j)}y' \right] + d$$

$$\geq \sum_{j=1}^{J} t_j(s_j - 1) + m - 1 - \sum_{l \in L}(s_l - 1) + d$$

$$\geq \sum_{j=1}^{J} t_j(s_j - 1) + m,$$

proving the validity of the induction hypothesis for this case.

The remaining case is if no nonsingular K-matrix T_j exists such that $T_j \Im_s(M^{(j)})$ has a row of the form $k[g_1 \, g_2 \, \cdots \, g_m]$, $g_i \in G$, $k \in K$, for $1 \leq j \leq J$. As for Theorem 3.5, a specialization of y_1 is guaranteed by Theorem 2.5 that will reduce the number of required m/d steps by at least one. The details will be omitted since this portion of the proof is identical to the corresponding part of the proof of Theorem 3.5. The final result is that the induction hypothesis must be satisfied for $(t, m-1)$, requiring at least one fewer m/d step than for (t, m) and

$$\mu_B(\underset{j=1}{\overset{J}{\cup}}(t_j A(f_j))M^{(j)}y) \geq \sum_{j=1}^{J} t_j(s_j - 1) + m - 1 + 1 = \sum_{j=1}^{J} t_j(s_j - 1) + m. \blacksquare$$

3.6. Products with Several Fixed Polynomials in Different Rings

As pointed out in §3.3, polynomial multiplication modulo a general polynomial can be reformulated by the Chinese remainder theorem into a set of polynomial multiplications modulo each of the irreducible factors of the original modulus polynomial. A system consisting of the union of polynomial products modulo several (possibly repeated) polynomials will be denoted by

$$\bigcup_{j=1}^{J} (t_j A_j(f_j)) M^{(j)} y.$$

This system will be considered quasi-row-column reduced if it satisfies the conditions specified in Definition 3.4.

Theorem 3.9. [6] *If* $\bigcup_{j=1}^{J} (t_j A_j(f_j)) M^{(j)} y$ *is a qrc reduced system with the cardinality of* y *equal to* m *and the row rank of* f_j *equal to* $s_j \geq 1$, $j = 1, 2, \ldots, J$, *and for every* $L \subseteq \{1, 2, \ldots, J\}$, $\dim_L G(\bigcup_{l \in L} r(f_l)) \geq \sum_{l \in L} (s_l - 1)$, *then*

$$\mu_B(\bigcup_{j=1}^{J} (t_j A_j(f_j)) M^{(j)} y; G) \geq \sum_{j=1}^{J} t_j (s_j - 1) + m,$$

where $r(f_l)$ *and* B *are as in the statement of Theorem 3.8.*

Proof. Let K_j be the field $G[u]/\langle P_j(u) \rangle$, where the P_j's are polynomials over G. The system $\bigcup_{j=1}^{J} (t_j A_j(f_j)) M^{(j)} y$ may also be denoted by $\bigcup_{j=1}^{J} \bigcup_{\alpha(j)=1}^{t_j} C(K_j; f_j, l_{\alpha(j)})$. Let K be a finite extension of G such that each $K_j \subseteq K$, $j = 1, 2, \ldots, J$. Equation (3.6) showed that $C(K_j; f_j, l_{\alpha(j)})$ is equivalent to $C(K; f_j', l_{\alpha(j)}')$ where f_j' and $l_{\alpha(j)}'$ are the images of f_j and $l_{\alpha(j)}$ in $F \otimes_G K$. This equivalence then implies that $\bigcup_{j=1}^{J} \bigcup_{\alpha(j)=1}^{t_j} C(K_j; f_j, l_{\alpha(j)})$ is equivalent to $\bigcup_{j=1}^{J} \bigcup_{\alpha(j)=1}^{t_j} C(K; f_j', l_{\alpha(j)}') = \bigcup_{j=1}^{J} t_j A(f_j') \overline{M}^{(j)} y$.

Based on the assumptions of this theorem and the construction of the field K, the system $\bigcup_{j=1}^{J} t_j A(f_j') \overline{M}^{(j)} y$ must clearly be qrc reduced and satisfy the other conditions of Theorem 3.8. Thus by Theorem 3.8,

$$\mu_B(\bigcup_{j=1}^{J} (t_j A_j(f_j)) M^{(j)} y; G) \geq \sum_{j=1}^{J} t_j (s_j - 1) + m. \quad \blacksquare$$

3.7. Multivariate Polynomial Multiplication

We have now determined the number of multiplications necessary to compute various types of polynomial products in one variable. A problem that arises when evaluating the multiplicative complexity of the discrete Fourier transform is that of determining the number of m/d steps necessary to compute a polynomial product in several variables modulo univariate polynomials in each of the variables. We will show that multivariate polynomial products are equivalent to univariate polynomial products, from which the multiplicative complexity can be deduced. The approach is first presented for two-variable systems, and the generalization to more variables is demonstrated by applying the approach to multidimensional cyclic convolution.

3.7.1. Polynomial Products in Two Variables

The systems that will be investigated in this section are those of the form

$$T(u, v) = R(u, v)S(u, v) \pmod{P(u), Q(v)} \qquad \text{where} \qquad R(u, v) = \sum_{j=0}^{q-1} \sum_{i=0}^{p-1} x_{ij} u^i v^j \qquad \text{and}$$

$S(u, v) = \sum_{j=0}^{q-1} \sum_{i=0}^{p-1} y_{ij} u^i v^j$ have indeterminate coefficients, and $P(u) = \sum_{i=0}^{p} g_i u^i$ and

$Q(v) = \sum_{i=0}^{d} g_i' v^i$ are irreducible polynomials over G. This system will be denoted by

$MPM(P, Q)$. Winograd [48] has investigated this system and has determined the multiplicative complexity of multivariate polynomial multiplication when the modulus polynomials have no repeated roots.

Theorem 3.10. [48] *If β is a root of Q and P factors over $G(\beta)$ into m irreducible polynomials, then $\mu_\beta(MPM(P, Q); G) = 2pq - m$.*

Proof. The system $MPM(P, Q)$ is equivalent to the tensor product $MPM(P) \otimes MPM(Q)$ since $R(u, v)$, $S(u, v)$, and $T(u, v)$ can be expressed as polynomials in v, each of whose coefficients is a polynomial in u, where multiplication of these coefficients implies polynomial multiplication followed by reduction modulo $P(u)$. A "general" element of $G(\beta)$ can be expressed as $x^{(i)} = \sum_{j=0}^{q-1} x_j^{(i)} \beta^j$. Computation of the coefficients of a polynomial product modulo $Q(v)$ is equivalent to multiplication of two "general" elements of $G(\beta)$, since the product of these two elements is a product of polynomials in β, and, because β is a root of $Q(v)$, the resultant product is reduced modulo $Q(\beta)$. The system $MPM(P, Q)$ can be expressed as the system $MPM(P)$ where each of the indeterminates has been replaced by a vector of indeterminates corresponding to the coefficients of each of the powers of β. Since P is reducible in $G(\beta)$, we obtain by the Chinese remainder theorem that $MPM(P, Q)$ is isomorphic to the direct sum $MPM(P_1) \oplus MPM(P_2) \oplus \cdots \oplus MPM(P_m)$ over $G(\beta)$. Each of the systems of polynomial multiplication modulo P_j can be replaced by a product of two "general" elements in the field $G(\beta, \alpha_j)$ where α_j is a root of P_j. This procedure can now be reversed by defining V_j to be a polynomial whose root generates $G(\beta, \alpha_j)$, thus each of the products in this extension field can be replaced by polynomial multiplication modulo V_j over G. The degree of V_j is $\deg P_j \cdot \deg Q$. Therefore $MPM(P, Q)$ is isomorphic to the direct sum $MPM(V_1) \oplus MPM(V_2) \oplus \cdots \oplus MPM(V_m)$ that requires $\sum_{i=1}^{m} (2 \deg P_i \deg Q - 1) = 2pq - m$ m/d steps over G by Theorem 3.3. ∎

This theorem only provides the number of multiplications when the polynomials P and Q have no repeated roots. This restriction causes no problems in the evaluation of the multiplicative complexity of either the DFT or multidimensional cyclic convolution since the modulus polynomials for these systems are cyclotomic polynomials that have no repeated roots, and none of the cyclotomic polynomials are repeated in a single variable.

Since this theorem is important in the analysis of the multiplicative complexity of the multidimensional DFT, two examples will be presented to demonstrate its application. The first example uses two polynomial moduli whose roots do not split each other. Even for this case the number of required m/d steps is less than that used by forming the tensor product of the two component systems. The second example presents a system in which each polynomial modulus can be factored over the roots of the other, leading to a further decrease in the multiplicative complexity.

Example 3.5. Consider the system $MPM(u^2+1, v^2+v+1)$ that computes the product $R(u, v)S(u, v) \pmod{u^2+1, v^2+v+1}$, where

$$R(u, v) = x_0 + x_1 u + x_2 v + x_3 uv,$$

$$S(u, v) = y_0 + y_1 u + y_2 v + y_3 uv$$

and $G = Q$, the field of rational numbers. The two modulus polynomials are the 4th and 3rd cyclotomic polynomials respectively, and neither can be factored over the rationals extended by a root of the other, as demonstrated in Theorem A.1. In the following w_i will denote the ith root of unity.

The resultant univariate polynomial multiplication will be modulo the polynomial generating the field $Q(w_3, w_4) = Q(w_{12})$, which is $C_{12}(t) = t^4 - t^2 + 1$. Each occurrence of u in R and S is replaced by $w_4 = w_{12}^3 = t^3$ and v is similarly replaced by $w_3 = w_{12}^4 = t^4$ yielding

$$R(t) = x_0 + x_1 t^3 + x_2 t^4 + x_3 t^7$$

$$\equiv (x_0 - x_2) - x_3 t + x_2 t^2 + x_1 t^3 \pmod{t^4 - t^2 + 1}$$

and

$$S(t) \equiv (y_0 - y_2) - y_3 t + y_2 t^2 + y_1 t^3 \pmod{t^4 - t^2 + 1}.$$

Theorem 3.2 states that seven multiplications are necessary to compute the product $R(t)S(t) \pmod{t^4 - t^2 + 1}$. After computing the product $R(t)S(t) = z_0 + z_1 t + z_2 t^2 + z_3 t^3$, the inverse mapping obtains

$$R(u,v) = (z_0+z_2)+z_3u+z_2v-z_1uv.$$

If an algorithm had been constructed for the system of Example 3.5 as a tensor product of minimal algorithms for the two components, then the number of m/d steps in the resulting algorithm would be the product of the multiplicative complexities of the two component systems. The two component systems require three m/d steps each resulting in a total of nine m/d steps for the tensor product algorithm. Therefore the construction of Theorem 3.10 uses two fewer m/d steps than the tensor product algorithm.

The disadvantage of using the minimal algorithm in Example 3.5 is that the product modulo t^4-t^2+1 requires seven distinct constants from Q, each of which is raised to the third power, resulting in rational multiplications that must be carried out with many additions. As the degrees of the component polynomial moduli in the tensor products get larger the number of required constants increases, as does the exponent to which each must be raised, resulting in impractical minimal algorithms. In contrast, the tensor product algorithm for Example 3.5 requires only three distinct constants from G, which may be used in both of the minimal algorithms for the component systems.

Winograd [47] suggests that the construction of Theorem 3.10 can be reversed in certain cases, resulting in tensor product algorithms for computing products modulo irreducible polynomials of large degree. This suggestion is only applicable when the root of the irreducible polynomial can be expressed as a product of roots of smaller irreducible polynomials over G.

Example 3.6. Consider the system $MPM(u^2+u+1, v^2-v+1)$ that computes $R(u,v)S(u,v) \pmod{u^2+u+1, v^2-v+1}$, where

$$R(u,v) = x_0+x_1u+x_2v+x_3uv,$$

$$S(u,v) = y_0+y_1u+y_2v+y_3uv,$$

and $G = Q$, the field of rational numbers. The modulus polynomials are the 3rd and 6th cyclotomic polynomials and Theorem A.1 shows that each has two factors over the rationals extended by a root of the other. These factorizations are

$$u^2+u+1 = (u-w_6^2)(u-w_6^4)$$

$$= (u-w_6+1)(u+w_6)$$

and

$$v^2-v+1 = (v+w_3)(v+w_3^2)$$

$$= (v+w_3)(v-w_3-1).$$

At this point either u or v can be replaced by the corresponding root. The number of m/d steps is the same either way, although the number of additions may vary for the two methods. We will proceed by replacing v with w_6. The reductions for $R(u, v)$ and $S(u, v)$ are identical, so only one will be shown,

$$R(u) = (x_0 + x_2 w_6) + (x_1 + x_3 w_6)u,$$

$$R(u) \equiv (x_0 + x_3) + (x_2 - x_1 - x_3)w_6 \pmod{u + w_6},$$

$$R(u) \equiv (x_0 - x_1 - x_3) + (x_2 + x_1)w_6 \pmod{u - w_6 + 1}.$$

The product $R(u)S(u)$ is computed for both of the factors of $u^2 + u + 1$ using minimal algorithms for multiplication modulo $v^2 - v + 1$, for a total of 6 m/d steps.

The two products are $R(u)S(u) \equiv z_0 + z_1 w_6 \pmod{u + w_6}$ and $R(u)S(u) \equiv z_0' + z_1' w_6 \pmod{u - w_6 + 1}$. The CRT reconstruction is

$$R(u)S(u) \equiv (z_0 + z_1 w_6)(u - w_6 + 1)\left\lceil \frac{2w_6 - 1}{3} \right\rceil + (z_0' + z_1' w_6)(u + w_6)\left\lceil \frac{1 - 2w_6}{3} \right\rceil$$

$$= \frac{1}{3}\left[[(z_0 - z_1 + 2z_0' + z_1') + (z_0 + 2z_1 - z_0' + z_1')w_6] + \right.$$

$$\left. + [(-z_0 - 2z_1 + z_0' + 2z_1') + (2z_0 + z_1 - 2z_0' - z_1')w_6]u \right]$$

yielding

$$R(u,v)S(u,v) = \frac{1}{3}\left[(z_0 - z_1 + 2z_0' + z_1') + (-z_0 - 2z_1 + z_0' + 2z_1')u + \right.$$

$$\left. + (z_0 + 2z_1 - z_0' + z_1')v + (2z_0 + z_1 - 2z_0' - z_1')uv \right]$$

when w_6 is replaced by v again.

The system of Example 3.6 decomposed into the direct sum of two polynomial products, allowing the same set of constants from Q to be used for both products. Therefore it may be advantageous to use minimal algorithms in computing multivariate polynomial products when the roots of the modulus polynomials are related in a way such that m in Theorem 3.10 is greater than one. This algorithm required three fewer m/d steps than the tensor product algorithm, demonstrating the added savings possible when a root of one polynomial splits the other polynomial.

The *polynomial transforms* proposed by Nussbaumer [27] yield algorithms identical to those derived in this section. Polynomial transforms provide an interesting and different perspective on these types of systems.

3.7.2. Multidimensional Cyclic Convolution

Theorem 3.10 can be extended to cases in which the number of variables is greater than two. This extension is obtained by iteratively applying Theorem 3.10 to each variable. The following corollary applies this technique to multidimensional cyclic convolution. The function ϕ is Euler's (totient) function, defined in Appendix A.

Corollary 3.2. $\mu_B(MPM(u_1^{N_1}-1, u_2^{N_2}-1, \ldots, u_m^{N_m}-1))$

$$= 2\prod_{i=1}^{m}N_i - \sum_{d_1 \mid N_1}\sum_{d_2 \mid N_2}\cdots\sum_{d_m \mid N_m}\frac{\prod_{i=1}^{m}\phi(d_i)}{\phi([d_1, d_2, \ldots, d_m])}.$$

Proof. Corollary 3.1 can be applied in each dimension of the multidimensional cyclic convolution to express the system $\bigotimes_{i=1}^{m} MPM(u_i^{N_i}-1)$ as $\bigotimes_{i=1}^{m} \bigoplus_{d_i \mid N_i} MPM(C_{d_i})$. Combining the result of Theorem A.1 of Appendix A with Theorem 3.10 shows that $MPM(C_{d_1}, C_{d_2})$ is equivalent to the direct sum $\bigoplus_{i=1}^{\phi((d_1, d_2))} MPM(C_{[d_1, d_2]})$, where C_d is the dth cyclotomic polynomial, (d_1, d_2) is the greatest common divisor of d_1 and d_2, and $[d_1, d_2]$ is the least common multiple of d_1 and d_2. Addition of a third dimension yields

$$MPM(C_{d_1}, C_{d_2}, C_{d_3}) = MPM(C_{d_1}, C_{d_2}) \otimes MPM(C_{d_3})$$

$$= \left[\bigcup_{i=1}^{\phi((d_1, d_2))} MPM(C_{[d_1, d_2]})\right] \otimes MPM(C_{d_3})$$

$$= \bigcup_{i=1}^{\phi((d_1, d_2))} \left[MPM(C_{[d_1, d_2]}) \otimes MPM(C_{d_3})\right]$$

$$= \bigcup_{i=1}^{\phi((d_1, d_2))} \left[\bigcup_{j=1}^{\phi(([d_1, d_2], d_3))} MPM(C_{[d_1, d_2, d_3]})\right]$$

$$= \bigcup_{i=1}^{\frac{\phi(d_1)\phi(d_2)\phi(d_3)}{\phi([d_1, d_2, d_3])}} MPM(C_{[d_1, d_2, d_3]}).$$

This construction can be repeated for additional dimensions, resulting in

$$\overset{m}{\underset{i=1}{\otimes}} MPM(C_{d_i}) = \overset{\frac{\prod\limits_{j=1}^{m}\phi(d_j)}{\phi([d_1,d_2,\ldots,d_m])}}{\underset{i=1}{\cup}} MPM(C_{[d_1,d_2,\ldots,d_m]}).$$

Returning to the problem of multidimensional cyclic convolution, we obtain

$$\overset{m}{\underset{i=1}{\otimes}} \underset{d_i|N_i}{\oplus} MPM(C_{d_i}) = \underset{d_1|N_1}{\oplus} \underset{d_2|N_2}{\oplus} \cdots \underset{d_m|N_m}{\oplus} \left[\overset{m}{\underset{i=1}{\otimes}} MPM(C_{d_i})\right]$$

$$= \underset{d_1|N_1}{\oplus} \underset{d_2|N_2}{\oplus} \cdots \underset{d_m|N_m}{\oplus} \left[\overset{\frac{\prod\limits_{j=1}^{m}\phi(d_j)}{\phi([d_1,d_2,\ldots,d_m])}}{\underset{i=1}{\cup}} MPM(C_{[d_1,d_2,\ldots,d_m]})\right].$$

This entire system satisfies the conditions of Theorem 3.9, therefore the multiplicative complexity can be computed as

$$\mu_B\left(\overset{m}{\underset{i=1}{\otimes}} MPM(u_i^{N_i}-1); G\right) = \underset{d_1|N_1}{\sum} \underset{d_2|N_2}{\sum} \cdots \underset{d_m|N_m}{\sum} \frac{(2\phi([d_1,d_2,\ldots,d_m])-1)\prod\limits_{i=1}^{m}\phi(d_i)}{\phi([d_1,d_2,\ldots,d_m])}$$

$$= \underset{d_1|N_1}{\sum} \underset{d_2|N_2}{\sum} \cdots \underset{d_m|N_m}{\sum} \left[2\overset{m}{\underset{i=1}{\prod}}\phi(d_i) - \frac{\prod\limits_{i=1}^{m}\phi(d_i)}{\phi([d_1,d_2,\ldots,d_m])}\right]$$

$$= 2\overset{m}{\underset{i=1}{\prod}}N_i - \underset{d_1|N_1}{\sum} \underset{d_2|N_2}{\sum} \cdots \underset{d_m|N_m}{\sum} \frac{\prod\limits_{i=1}^{m}\phi(d_i)}{\phi([d_1,d_2,\ldots,d_m])}. \quad\blacksquare$$

The multiplicative complexity of multidimensional cyclic convolutions is tabulated in Table B.1 in Appendix B for all such systems requiring fewer than 100 m/d steps. This table suggests that multidimensional cyclic convolutions are equivalent to one-dimensional cyclic convolutions when the length in each dimension is relatively prime to the lengths in every other dimension. This is a trivial corollary of Theorem 3.10 and Corollary 3.2 since each cyclotomic polynomial modulus suggested by Corollary 3.2 is distinct and together they comprise the set of divisors of the total number of points.

A good example of this is for 30 total points whose one-dimensional multiplicative complexity from Corollary 3.1 is $2\cdot30-\tau(30) = 60-8 = 52$ m/d steps. From Table B.1 it can be seen that multidimensional cyclic convolutions of lengths $2\times3\times5$,

2×15, 3×10, and 5×6 also require 52 m/d steps each. This equivalence between the one-dimensional and multidimensional cyclic convolutions is exactly the mapping discovered by Agarwal and Cooley [1] and used in deriving the *rectangular transform* algorithms. The following example demonstrates this conversion of multidimensional cyclic convolution into one-dimensional cyclic convolution.

Example 3.7. Consider the system $MPM(u^3-1, v^2-1)$ as the product $R(u,v)S(u,v) \pmod{u^3-1, v^2-1}$ where

$$R(u,v) = x_0 + x_1 u + x_2 u^2 + x_3 v + x_4 uv + x_5 u^2 v$$

and

$$S(u,v) = y_0 + y_1 u + y_2 u^2 + y_3 v + y_4 uv + y_5 u^2 v.$$

This system can be expressed as

$$MPM(u^3-1, v^2-1) = MPM(C_3 C_1) \otimes MPM(C_2 C_1)$$
$$= \left[MPM(C_3) \oplus MPM(C_1)\right] \otimes \left[MPM(C_2) \oplus MPM(C_1)\right]$$
$$= \left[MPM(C_3) \otimes MPM(C_2)\right] \oplus \left[MPM(C_3) \otimes MPM(C_1)\right] \oplus$$
$$\left[MPM(C_1) \otimes MPM(C_2)\right] \oplus \left[MPM(C_1) \otimes MPM(C_1)\right]$$
$$= MPM(C_6) \oplus MPM(C_3) \oplus MPM(C_2) \oplus MPM(C_1)$$
$$= MPM(C_6 C_3 C_2 C_1)$$
$$= MPM(u^6-1).$$

Replacing u with $w_3 = t^2$ and v with $w_2 = t^3$ yields the system $R(t)S(t) \pmod{t^6-1}$ with

$$R(t) = x_0 + x_5 t + x_1 t^2 + x_3 t^3 + x_2 t^4 + x_4 t^5,$$

$$S(t) = y_0 + y_5 t + y_1 t^2 + y_3 t^3 + y_2 t^4 + y_4 t^5,$$

and a similar correspondence between the coefficients of $R(u,v)S(u,v)$ and $R(t)S(t)$, when each is reduced by its modulus polynomial(s).

3.8. Summary of Chapter 3

This chapter began by presenting an analysis of the multiplicative complexity of the product of an mth and n^{th} degree polynomial, showing that $\mu_B(PM(m,n); G) = m+n+1$. It was shown that when two polynomials of degree $N-1$ are multiplied and reduced modulo (a power of) an irreducible polynomial of degree N, then the same number of m/d steps (i.e., $m+n+1 = 2N-1$) are needed, irrespective of the residue reduction. If the modulus polynomial factors over G and has k distinct irreducible factors, then the required number of m/d steps was shown to be $2N-k$. This result follows from the Chinese remainder theorem decomposition of this system into a direct sum of polynomial products modulo each of the irreducible factors.

An algebraic setting was then formulated to permit the earlier results to be extended to semilinear systems and also to unions of these systems for which one or more modulus polynomials and/or the fixed polynomials in the component products are repeated. Under suitable conditions it was shown that the multiplicative complexity of the aggregate system is equal to the sum of the complexities of the component systems in the direct sum. When a component system is incomplete (i.e., one of the polynomials does not contain as many indeterminates as the degree of the modulus polynomial), the lower bound on the multiplicative complexity is reduced by one for each "missing" indeterminate as long as each polynomial has at least one indeterminate represented in its coefficients.

The last part of this chapter showed that a product of multivariate polynomials modulo univariate polynomials in each variable is equivalent to a direct sum of polynomial products modulo univariate polynomials and can thus be analyzed using the methods discovered earlier in the chapter. Multidimensional cyclic convolution was considered as an example of this type of system and the multiplicative complexity was determined for all possible dimensionalities and lengths.

CHAPTER 4

Constrained Polynomial Multiplication

Frequently, one or both of the polynomials to be multiplied has some special characteristic, such as symmetry in the coefficients, coefficients that are zero or in G, or some other possibility. What are the effects of input constraints on the multiplicative complexity of systems of polynomial multiplication? This chapter shows precisely what types of constraints reduce multiplicative complexity. These results are then applied to the special case of products of polynomials exhibiting symmetry.

A second type of constraint on a system of polynomial multiplication occurs when one or more of the outputs need not be computed. We refer to this type of system as a polynomial product with restricted outputs. Although seemingly closely related to the systems with input constraints, this type of system is much more difficult to analyze, and no general method can be used to determine whether a specific system requires fewer m/d steps than are needed to compute the entire set of output coefficients. Decimated and truncated sets of outputs are considered as examples in this analysis.

4.1. General Input Constraints

Suppose that the coefficients of $x(u)$ (i.e., the entries of Φ) are not linearly independent over G, but the coefficients of $y(u)$ remain indeterminate. Certain linear constraints of this type clearly reduce the multiplicative complexity of the system Φy. For example, if the coefficients of $x(u)$ are all identical, then the number of m/d steps necessary to compute Φy is $n+1$. An easy way to see this is to express $x(u)$ as $x_0 \cdot (\sum_{i=0}^{m} u^i)$, from which $x(u)y(u)$ can be computed by first forming the product $x_0 y(u)$ using $n+1$ m/d steps, then multiplying by $\sum_{i=0}^{m} u^i$, requiring only non-m/d steps. Clearly any polynomial $x(u)$ that has a factor in $G[u]$ of degree $p \geq 1$ can be reduced to an equivalent polynomial of degree $m-p$ in analyzing the multiplicative complexity of products with $x(u)$.

There are other types of constraints for which the multiplicative complexity is reduced. If some polynomial in $G[u]$ can be added to $x(u)$ so that the resulting polynomial has a non-constant factor in $G[u]$, then the multiplicative complexity is reduced by the degree of this factor. This is simply the application of the homomorphism r that was defined in Chapter 3 to take care of this possibility. In this chapter we will not use the mapping r, but will account for such additive terms explicitly. The following lemma will be useful in converting inhomogeneous systems of equations caused by these additive terms into homogeneous systems of equations.

Lemma 4.1. *Let* $\Gamma x = d$, *where* Γ *is an* $n{+}1 \times m{+}1$ *nonzero* G-*matrix*, $x = [x_0\, x_1\, \cdots\, x_m]^T \notin G^{m+1}$, *and* $d = [d_0\, d_1\, \cdots\, d_n]^T \in G^{n+1}$. *The vector* x *can always be replaced by* $\bar{x}{+}e$, $e \in G^{m+1}$, *satisfying* $\Gamma \bar{x} = 0$ *and* $\Gamma e = d$.

Proof. Assume first that $m = n$ and rank $\Gamma = m{+}1$. Then Γ must be invertible and $x = \Gamma^{-1}d$, implying that $x \in G^{m+1}$, contradicting the assumption of the lemma. Similarly, if $m < n$ and rank $\Gamma = m{+}1$, then an $m{+}1 \times m{+}1$ submatrix of Γ is invertible, again forcing $x \in G^{m+1}$. Therefore when $m \leq n$ as well as for $m > n$, rank $\Gamma < m{+}1$ and the system $\Gamma e = d$ is underdetermined, yielding infinitely many solutions $e \in G^{m+1}$, any of which may be chosen as $\bar{x} = x{-}e$ satisfying the statement of the lemma. ■

When x is the vector of coefficients of $x(u)$ and $\rho_r(\Phi) < m{+}n$, this lemma states that a polynomial $Q(u) \in G[u]$ of degree less than or equal to m can be added to $x(u)$ to convert the constraining equations into a homogeneous set of equations. If the equations were homogeneous originally, then $Q(u)$ could be zero, or since infinitely many solutions exist it could be something else. A simple example will illustrate the application of Lemma 4.1.

Example 4.1. Suppose the system $PM(1, 1)$ is given and the coefficients of $x(u)$ satisfy $\Gamma x = d$ where $\Gamma = \begin{bmatrix} c_0 & c_1 \\ c_1 & c_2 \end{bmatrix}$, $x = \begin{bmatrix} x_0 \\ x_1 \end{bmatrix} \notin G^2$, and $d = \begin{bmatrix} d_0 \\ d_1 \end{bmatrix} \in G^2$, and the entries of Γ are elements of G, at least one of which is nonzero. A vector $e = [e_0\, e_1]^T \in G^2$ must be selected satisfying $\Gamma e = d$.

The matrix Γ is singular, as pointed out in the lemma, and a simple argument can be used to show that if any one of c_0, c_1, c_2 is nonzero then all three are nonzero (this saves us from worrying about division by zero). The equation $c_0 e_0 {+} c_1 e_1 = d_0$ yields $e_1 = \dfrac{-c_0 e_0}{c_1} + \dfrac{d_0}{c_1}$ where $e_0 \in G$ may be selected arbitrarily. Now $\bar{x} = x{-}e$ must satisfy $\Gamma \bar{x} = 0$ and the polynomial $x(u)$ may be replaced by the polynomial $x(u){+}e_0{+}e_1 u$ without affecting $\rho_r(\Phi)$.

The basic idea used in the proof of Lemma 3.1 is that $\rho_r(\Phi)$, the row rank over G of the matrix Φ containing the coefficients of $x(u)$, is equal to $m{+}n{+}1$, the number of rows in Φ, since each of the coefficients is a distinct indeterminate. The following theorem states exactly when $\rho_r(\Phi)$ is less than $m{+}n{+}1$ for n equal to or larger than m.

Theorem 4.1. *Let* $y(u) = \sum_{i=0}^{n} y_i u^i$ *be a polynomial with distinct coefficients in B that are linearly independent over* G, *and let* $x(u) = \sum_{i=0}^{m} x_i u^i$ *be a polynomial whose coefficients are in B. Let* $n \geq m$. *If Z is the system defined by the product* $x(u)y(u)$, *then* $\mu_B(Z; G) = m+n+1$, *unless* $x(u)$ *can be expressed as* $x(u) = Q(u) + \sum_{i=0}^{m'} x_i' u^i Q'(u)$ *where* $Q(u), Q'(u) \in G[u]$ *and* $m' < m$, *in which case* $\mu_B(Z; G) = m'+n+1$.

Proof. The entries of Φ are $\phi_{ij} = x_{i-j}$, $j = 0, 1, \ldots, n$ and i such that $j \leq i \leq j+m$, and $\phi_{ij} = 0$ otherwise (rows and columns are numbered from zero to better identify entries as polynomial coefficients). If $\rho_r(\Phi) < m+n+1$, then there must be a nonzero row vector $c = [c_0 \, c_1 \, \cdots \, c_{m+n}]$ with entries in G such that $c\Phi = d \in G^{n+1}$. As was shown in Lemma 4.1, the polynomial $x(u)$ can be replaced by $x(u)+Q(u)$, where $Q(u) \in G[u]$, such that for the new system $c\Phi = 0$. For the rest of this proof it will be assumed that the coefficients of $x(u)$ satisfy the equation $c\Phi = 0$, since they could always be converted to such a system without altering $\rho_r(\Phi)$. Another assumption will be that $x_m \notin G$, since otherwise the system is equivalent to one with an $x(u)$ of degree less than m.

The inner product of c with the ith column of Φ gives the constraining equation $\sum_{j=0}^{m} c_{j+i} x_j = 0$, $i = 0, 1, \ldots, n$. This can be expressed as the product $\Gamma x = 0$ where $\gamma_{ij} = c_{i+j}$ are the entries in the $n+1 \times m+1$ matrix Γ and $x = [x_0 \, x_1 \, \cdots \, x_m]^T$ is the vector of coefficients of $R(u)$. Each linearly independent row of

$$\Gamma = \begin{bmatrix} c_0 & c_1 & \cdots & c_m \\ c_1 & c_2 & \cdots & c_{m+1} \\ \vdots & \vdots & \ddots & \vdots \\ c_n & c_{n+1} & \cdots & c_{m+n} \end{bmatrix}$$

implies an independent linear relationship between the rows of Φ, and hence reduces $\rho_r(\Phi)$ by one. Let p be the maximum number of linearly independent rows of Γ over G. When $m \leq n$, then p is equal to the dimension of the null space of Φ^T over G, and also $\rho_r(\Phi) = m+n+1-p$. When $m > n$ then Γ has only $n+1$ rows and it is possible for the dimension of the null space of Φ^T over G to be greater than the number of independent rows of Γ.

Consider the case when $m \leq n$. Suppose that the first $p-q$ rows of Γ (numbered zero through $p-q-1$) are linearly independent over G and the $(p-q)^{th}$ row is dependent on the previous $p-q$ rows. From this dependence, the columns of Γ provide the

relations $\quad c_i = \sum_{j=0}^{p-q-1} g_j c_{j+i-p+q}, \quad i = p-q, p-q+1, \ldots, m+p-q, \quad$ where $\quad g_j \in G,$
$j = 0, 1, \ldots, p-q-1$. We will show by induction that each c_i, $p-q \le i \le m+n$, must satisfy the difference equation $c_i = \sum_{j=0}^{p-q-1} g_j c_{j+i-p+q}$.

By assumption, c_{p-q} satisfies the difference equation, thus suppose that each c_i satisfies the equation for $i < s$. The first row in which c_s appears is the $(s-m)^{th}$ row, as the entry in the final column. Since each of the other entries in the $(s-m)^{th}$ row satisfy the difference equation, a linear combination of the previous $p-q$ rows can be formed, which when subtracted from the $(s-m)^{th}$ row yields a row that is nonzero only in the final column. The inner product of this row with x must be 0, therefore $(c_s - \sum_{j=0}^{p-q-1} g_j c_{j+s-p+q}) x_m = 0$, but since c_s does not satisfy the difference equation, the parenthesized expression can not be zero and thus $x_m = 0$. This violates the statement of the theorem by causing the polynomial $x(u)$ to be of degree less than m, therefore $c_s = \sum_{j=0}^{p-q-1} g_j c_{j+s-p+q}$, and by induction on s, this equation must be satisfied for each c_i, $p-q \le i \le m+n$. Every row of Γ is dependent on the first $p-q$ rows, therefore q must be zero if p rows are to be independent.

The difference equation governing the entries of Γ can be converted into a weighted sum of exponential terms based on the roots of the z-transform of the difference equation coefficients. It will first be assumed that all roots of this polynomial are distinct, so that

$$c_i = \sum_{k=0}^{p-1} \alpha_k r_k^i; \quad i = 0, 1, \ldots, m+n, \tag{4.1}$$

is the generating function for the c_i, with r_k, $k = 0, 1, \ldots, p-1$, the p distinct roots of the difference equation. The matrix equation $\Gamma x = 0$ is equivalent to the inner products

$$\sum_{j=0}^{m} c_{j+i} x_j = 0; \quad i = 0, 1, \ldots, n,$$

which give

$$\sum_{j=0}^{m} \sum_{k=0}^{p-1} \alpha_k r_k^{i+j} x_j = 0; \quad i = 0, 1, \ldots, n, \tag{4.2}$$

when c_i is substituted from (4.1). After rearranging terms the following set of equations is obtained,

$$\sum_{k=0}^{p-1} \alpha_k r_k^i \left(\sum_{j=0}^{m} x_j r_k^j \right) = 0; \quad i = 0, 1, \ldots, n. \tag{4.3}$$

Let v be the vector whose entries are $v_i = \sum_{j=0}^{m} x_j r_k^j$, $k = 0, 1, \ldots, p-1$. Let D be the matrix whose entries are $d_{ik} = \alpha_k r_k^i$, $i = 0, 1, \ldots, n$, $k = 0, 1, \ldots, p-1$. The matrix formulation of Equation (4.3) is $Dv = 0$. The r_i are all distinct, $\alpha_k \neq 0$ for $k = 0, 1, \ldots, p-1$, and $p-1 \leq n$, therefore the matrix D has column rank p and p rows of D can be chosen that are linearly independent. These rows span a p-dimensional space, and thus the inner product of each of these rows with a nonzero p-vector cannot all be zero, implying that the vector v must be a vector of zeros. The ith entry of v is the polynomial $\sum_{j=0}^{m} x_j u^j$ evaluated at the ith root of the difference equation, and since this polynomial is zero at each of these roots, it must be divisible by $(u-r_i)$, $i = 0, 1, \ldots, p-1$, and thus by their product $\prod_{i=0}^{p-1}(u-r_i)$, which is the rational polynomial characterizing the difference equation.

When the difference equation has repeated roots, let q be the number of distinct roots and p_k the multiplicity of the kth root, then Equation (4.1) can be reformulated as

$$c_i = \sum_{k=0}^{q-1} \sum_{h=0}^{p_k-1} \alpha_{kh} i^h r_k^i; \quad i = 0, 1, \ldots, m+n,$$

yielding

$$\sum_{j=0}^{m} \sum_{k=0}^{q-1} \sum_{h=0}^{p_k-1} \alpha_{kh}(i+j)^h r_k^{i+j} x_j = 0; \quad i = 0, 1, \ldots, n,$$

in place of (4.2). A preliminary regrouping is

$$\sum_{j=0}^{m} \sum_{k=0}^{q-1} r_k^{i+j} x_j \sum_{h=0}^{p_k-1} \alpha_{kh}(i+j)^h = 0; \quad i = 0, 1, \ldots, n.$$

If the binomial term $(i+j)^h$ is expanded for each h in the innermost sum, the result can be thought of as a polynomial in j with each of the coefficients a polynomial in i. Each α_{kh} is replaced by α'_{ikh} such that

$$\sum_{h=0}^{p_k-1} \alpha_{kh}(i+j)^h = \sum_{h=0}^{p_k-1} \alpha'_{ikh} j^h; \quad k = 0, 1, \ldots, q-1, \quad i = 0, 1, \ldots, n,$$

by incorporating all dependence on i into the α' coefficients.

The equation $Dv = 0$ can again be generated, and again the columns of D are linearly independent, implying that v is a zero vector. Here some of the entries of v

are of the form $\sum_{j=0}^{m} x_j{}^h r_k^j$, which is zero for $h = 0, 1, \ldots, p_k-1$, constituting p_k

independent equations that are true only when r_k is a root of $\sum_{j=0}^{m} x_j u^j$ of multiplicity

p_k. This again implies that each root of the difference equation is also a root of $\sum_{j=0}^{m} x_j u^j$, and thus that the polynomial characterizing the difference equation divides $\sum_{j=0}^{m} x_j u^j$. ∎

This result is very gratifying since it is exactly equal to the upper bound that was derived earlier. Unfortunately, the analysis of this type of system for $m > n$ is more complicated. An upper bound on the multiplicative complexity for $m > n$ can be stated as a simple extension to the result for $m \le n$. Consider, for example, the system $(x_0 + x_6 u^6 + x_7 u^7)(y_0 + y_1 u)$. The polynomial $x(u)$ has no non-constant factors with coefficients in G, but clearly $\rho_r(\Phi) = 5$ and this system can be computed in two parts as $x_0(y_0 + y_1 u)$ and $(x_6 + x_7 u)(y_0 + y_1 u)u^6$ and summed, for a total of $2+3 = 5$ m/d steps.

This example suggests that if $x(u)$ can be expressed as $Q(u) + \sum_{j=0}^{r} \sum_{i=0}^{m_j} x'_{ij} u^i Q_i(u)$

where $Q(u), Q_i(u) \in G[u]$, $i = 1, 2, \ldots, r$, and if $\sum_{j=1}^{r} m_j < m - (n+1)(r-1)$, then

$\mu_B(xy; G) \le rn + r + \sum_{j=1}^{r} m_j < m + n + 1$. We have not proven that this is also a lower

bound on the multiplicative complexity.

4.2. Multiplication by a Symmetric Polynomial

Sometimes one or both of the polynomials to be multiplied has coefficients that exhibit some type of symmetry. The coefficients of linear phase finite impulse response (FIR) filters exhibit even symmetry about the center of the impulse response [28]. Therefore, the equivalent polynomial product has a multiplicand $x(u) = \sum_{i=0}^{m} x_i u^i$

with $x_i = x_{m-i}$, $i = 0, 1, \ldots, m$. This type of filter is normally implemented by initially adding each pair of inputs in the symmetric locations, giving a 50% savings in multiplications over the straightforward implementation.

The following result on the multiplicative complexity of multiplication by a symmetric polynomial seems strange since there is not a major reduction in multiplicative complexity over a general polynomial multiplication. This behavior can best be understood by examining the distinctive nature of symmetric polynomials.

Every root of a symmetric polynomial, $x(u)$, is clearly also a root of $x(u^{-1})$ since $x(u) = u^m x(u^{-1})$, and therefore the reciprocal of every root of $x(u)$ must also be a root of $x(u)$ with the same multiplicity. When $x(u)$ is of odd degree, then the number of roots is also odd, and therefore after pairing each root with its reciprocal, there will be one root left over that must be its own reciprocal. This root must satisfy $r = 1/r$, or $r^2 - 1 = 0$, yielding the possibilities $r = 1$ or $r = -1$. Clearly $r = -1$ is a root of all symmetric polynomials of odd degree since $x(-1) = \sum_{i=0}^{(m-1)/2} (x_i(-1)^i + x_i(-1)^{m-i}) = 0$. It is possible for $r = 1$ to be a root of a symmetric polynomial (e.g. $u^3 - u^2 - u + 1$), but its multiplicity must be an even number, and would imply linear relationships between the coefficients that are not characteristic of symmetric polynomials in general.

If $x(u)$ is a symmetric polynomial, the roots are reciprocal pairs of arbitrary complex numbers, except the root at -1 when the degree of $x(u)$ is odd. Therefore, in general, no rational polynomials divide $x(u)$ when the degree is even, and only one rational polynomial, $u+1$, divides $x(u)$ when the degree is odd. A product of a symmetric polynomial of degree m by a general polynomial of degree n will be denoted by $PM_s(m, n)$.

Theorem 4.2. *If $n > 0$, then $\mu_B(PM_s(m, n)) = m+n+1$ for m even, and $\mu_B(PM_s(m, n)) = m+n$ for m odd.*

Proof. The remarks of the previous paragraph show that the row rank over G of the matrix formed from $x(u)$ is $m+n+1$ when $\deg x$ is even and is $m+n$ when $\deg x$ is odd. Using Theorem 2.4, these constitute lower bounds on $\mu_B(PM_s(m, n))$ for m even or odd. ∎

4.3. Multiplication by an Antisymmetric Polynomial

When the polynomial $x(u)$ exhibits odd symmetry it is called antisymmetric or skew-symmetric. As for symmetric polynomials, there is a simple relation between the value of an antisymmetric polynomial at any nonzero point and the value at the reciprocal of this point. This relation is $x(u) = -u^m x(1/u)$, implying again that all roots (except for 1 and -1) must occur in reciprocal pairs. It is easily seen that 1 is a root of all antisymmetric polynomials, and when a single factor of $u-1$ is factored out of an antisymmetric polynomial, then the resulting polynomial must be a symmetric polynomial. The product of an antisymmetric polynomial of degree m with a general polynomial of degree n will be denoted by $PM_a(m, n)$.

Theorem 4.3. *If $n > 0$, then $\mu_B(PM_a(m, n)) = m+n-1$ for m even, and $\mu_B(PM_a(m, n)) = m+n$ for m odd.*

Proof. The factor of $u-1$ can be eliminated from the antisymmetric polynomial, leaving a system equivalent to $PM_s(m-1, n)$, and the result of Theorem 4.2 applied to obtain the result. ∎

Looking back at the proof of Theorem 4.2, we see that an antisymmetric polynomial of even degree has a factor of $u^2 - 1$ and an antisymmetric polynomial of odd degree has a factor of $u-1$. It is worth noting that antisymmetric polynomials of even degree must have a middle coefficient equal to zero. This is because each coefficient must satisfy $x_i = -x_{m-i}$, and when $i = m/2$, then $x_{m/2} = -x_{m/2}$, implying that $x_{m/2} = 0$.

4.4. Products of Two Symmetric Polynomials

Sometimes both of the polynomials to be multiplied are symmetric. When this is the case, the coefficients of $y(u)$ are no longer indeterminate and the row rank of y is $(n+1)/2$ for n odd, and $(n+2)/2$ for n even. The redundant rows can be eliminated by replacing each of the first $(n+1)/2$ or $n/2$ columns of Φ by the sum of the original column with its symmetric counterpart and eliminating the symmetric columns. Since $x(u)$ is symmetric, this "folding over" of the columns of Φ causes the rows of the resulting matrix to be symmetric, resulting in $(m+n+1)/2$ distinct rows when $m+n$ is odd, and $(m+n+2)/2$ distinct rows when $m+n$ is even. Because the columns were folded over, this system is, in general, no longer equivalent to an unrestricted polynomial multiplication and therefore the result of Theorem 4.1 is not applicable.

An upper bound on the number of m/d steps necessary to compute this system can be easily established by application of the Karatsuba-Toom algorithm and use of the following identities. Assume that m and n are both even, since if either or both were odd then factors of $u+1$ could be divided out where necessary. For any $g \in G$, $x(g) = g^m x(1/g)$, $y(g) = g^n y(1/g)$, and $T(g) = x(g)y(g) = g^{m+n}T(1/g)$. Therefore, for each $g \in G$, except for 1 and -1, two of the residues necessary for the Chinese remainder theorem reconstruction can be determined with only one m/d step. Since m and n were assumed to be even, the product is evaluated for a total of $m+n+1$ distinct elements of G, using a total of $\frac{m+n}{2}+1$ m/d steps. If either m or n is odd and the other even then $\frac{m+n-1}{2}+1$ m/d steps are sufficient, and if both m and n are odd then $\frac{m+n}{2}$ m/d steps are sufficient.

The number of m/d steps sufficient to compute the product $x(u)y(u)$ will now be shown to be the number of necessary m/d steps also. First, note that any general polynomial can be expressed as the sum of a symmetric and an antisymmetric polynomial of the same degree as the original polynomial. That is, for any m^{th} degree polynomial $x(u)$, the symmetric part is $x_s(u) = \frac{1}{2}[x(u)+u^m x(1/u)]$, and the antisymmetric part is $x_a(u) = \frac{1}{2}[x(u)-u^m x(1/u)]$, yielding $x(u) = x_s(u)+x_a(u)$. The product of two symmetric polynomials will be denoted by $PM_{ss}(m,n)$ and the product of a symmetric polynomial of degree m with an antisymmetric polynomial of degree n will be denoted by $PM_{sa}(m,n)$. The inequality $\mu_B(PM_s(m,n);G) \le \mu_B(PM_{ss}(m,n);G)+\mu_B(PM_{sa}(m,n);G)$ arises because an algorithm for computing $PM_s(m,n)$ could first decompose $y(u)$ into symmetric and antisymmetric components with no m/d steps, then algorithms for $PM_{ss}(m,n)$ and $PM_{sa}(m,n)$ could be implemented using the sum of the number of m/d steps required by each component, and finally the two component outputs could be added to obtain the symmetric polynomial multiplication.

When m and n are both even, then $\mu_B(PM_s(m, n); G) = m+n+1$, $\mu_B(PM_{ss}(m, n); G) \le \frac{m+n}{2}+1$, and $\mu_B(PM_{sa}(m, n); G) = \mu_B(PM_{ss}(m, n-2); G) \le \frac{m+n}{2}$. Adding the two inequalities yields $\mu_B(PM_{ss}(m, n); G)+\mu_B(PM_{sa}(m, n); G) \le m+n+1$, but also from the previous paragraph $\mu_B(PM_{ss}(m, n); G)$ $+\mu_B(PM_{sa}(m, n); G) \ge \mu_B(PM_s(m, n); G) = m+n+1$; therefore for the inequality to hold both ways, $\mu_B(PM_{ss}(m, n); G)+\mu_B(PM_{sa}(m, n); G) = m+n+1$. If either $\mu_B(PM_{sa}(m, n); G) < \frac{m+n}{2}$ or $\mu_B(PM_{ss}(m, n); G) < \frac{m+n}{2}+1$ then this equality could not be true, thus $\mu_B(PM_{ss}(m, n); G) = \frac{m+n}{2}+1$ and $\mu_B(PM_{sa}(m, n); G) = \frac{m+n}{2}$. The results for other values of m and n reduce to the case of m and n both even, and therefore a derivation of these results is omitted. Table 4.1 summarizes the multiplicative complexity of polynomial multiplication with symmetric and antisymmetric polynomials.

4.5. Polynomial Multiplication with Restricted Outputs

4.5.1. Decimation of Outputs

Frequently, only certain coefficients resulting from a product of polynomials are desired. It is clear, for instance, that if the coefficient of the constant term is the only required output of a system, then the system is equivalent to a product of two zero-order polynomials and one multiplication is necessary to compute the system. Computation of only the highest-order term can be similarly shown to require just a single multiplication. We have seen in the previous section that for linear constraints on the inputs to a system of polynomial multiplication it is possible to determine exactly what conditions must be satisfied for a constraint of this type to have an effect on the multiplicative complexity of the system. Can a similar statement be made regarding the effect of restricting the set of desired outputs to a subset of the coefficients resulting from a polynomial product?

As a first example, consider the computation of every d^{th} coefficient of a polynomial product. In digital filtering this selection of equally spaced output samples is called decimation, and the computation of these decimated outputs for a single polynomial product is a basic operation in the overlap-add method of digital filtering applied to a decimating filter. A similar problem has been analyzed by Winograd [47], but his analysis was confined to a technique more closely resembling overlap-save than overlap-add.

$x(u)$	$y(u)$	Parity of m	Parity of n	$\mu_B(PM_*(m,n))$
General	General	either	either	$m+n+1$
Symmetric	General	even	either	$m+n+1$
Symmetric	General	odd	either	$m+n$
Antisymmetric	General	even	either	$m+n-1$
Antisymmetric	General	odd	either	$m+n$
Symmetric	Symmetric	even	even	$\dfrac{m+n+2}{2}$
Symmetric	Symmetric	even	odd	$\dfrac{m+n+1}{2}$
Symmetric	Symmetric	odd	odd	$\dfrac{m+n}{2}$
Symmetric	Antisymmetric	even	even	$\dfrac{m+n}{2}$
Symmetric	Antisymmetric	even	odd	$\dfrac{m+n+1}{2}$
Symmetric	Antisymmetric	odd	even	$\dfrac{m+n-1}{2}$
Symmetric	Antisymmetric	odd	odd	$\dfrac{m+n}{2}$
Antisymmetric	Antisymmetric	even	even	$\dfrac{m+n-2}{2}$
Antisymmetric	Antisymmetric	even	odd	$\dfrac{m+n-1}{2}$
Antisymmetric	Antisymmetric	odd	odd	$\dfrac{m+n}{2}$

Table 4.1. Multiplicative complexity of computing $x(u)y(u)$
for symmetric and antisymmetric input polynomials.

Given a polynomial $x(u)$ of degree m and a polynomial $y(u)$ of degree n, the computation of every d^{th} coefficient will be denoted by $PM(m, n, d)$. Another parameter could be added to show which coefficient begins the decimated output, but this starting point does not affect the multiplicative complexity, so the same notation will be used to represent a class of systems with arbitrary starting points (less than d). It will be assumed that $d \leq m$ and $d \leq n$, since if this is not the case, the system reduces to the computation of disjoint inner products of vectors with the number of required multiplications equal to the column rank of the resulting matrix.

The system $PM(4, 4, 2)$ (with initial point 0) is represented in matrix notation as

$$
\begin{bmatrix} z_0 \\ z_2 \\ z_4 \\ z_6 \\ z_8 \end{bmatrix} = \begin{bmatrix} x_0 & 0 & 0 & 0 & 0 \\ x_2 & x_1 & x_0 & 0 & 0 \\ x_4 & x_3 & x_2 & x_1 & x_0 \\ 0 & 0 & x_4 & x_3 & x_2 \\ 0 & 0 & 0 & 0 & x_4 \end{bmatrix} \begin{bmatrix} y_0 \\ y_1 \\ y_2 \\ y_3 \\ y_4 \end{bmatrix}.
$$

Because of the 2:1 decimation every row of the above matrix is identical to the previous row shifted 2 columns to the right with new values in the leftmost 2 columns, thus each column of the matrix must contain coefficients of $x(u)$ whose indexes are all congruent (mod 2). Therefore the system $PM(4, 4, 2)$ can be expressed as the sum of $PM(2, 2)$ and $PM(1, 1)$ in the form

$$
\begin{bmatrix} z_0 \\ z_2 \\ z_4 \\ z_6 \\ z_8 \end{bmatrix} = \begin{bmatrix} x_0 & 0 & 0 \\ x_2 & x_0 & 0 \\ x_4 & x_2 & x_0 \\ 0 & x_4 & x_2 \\ 0 & 0 & x_4 \end{bmatrix} \begin{bmatrix} y_0 \\ y_2 \\ y_4 \end{bmatrix} + \begin{bmatrix} 0 & 0 \\ x_1 & 0 \\ x_3 & x_1 \\ 0 & x_3 \\ 0 & 0 \end{bmatrix} \begin{bmatrix} y_1 \\ y_3 \end{bmatrix}.
$$

When the decimation is $d:1$, the columns can be similarly permuted such that d partitions are formed, each consisting of columns whose original column indexes are congruent (mod d).

Theorem 4.4. *If $d \le m$ and $d \le n$ then $\mu_B(PM(m, n, d); G) = m+n-d+2$.*

Proof. Given a system $PM(m, n, d)$, it is always possible to express this system as a sum of disjoint systems in the manner shown in the previous example. Each of the component systems will be denoted by a subscript corresponding to the integer between zero and d to which the subscripts of the x_i's in that system are congruent (mod d). The ith system is equivalent to the system $PM(m_i, n_i)$, $i = 0, 1, \ldots, d-1$. Therefore since the systems are disjoint the total number of required multiplications is the sum of the number of multiplications required by each of these disjoint systems,

$$\mu_B(PM(m, n, d); G) = \sum_{i=0}^{d-1} \mu_B(PM(m_i, n_i); G)$$

$$= \sum_{i=0}^{d-1} (m_i + n_i + 1) \tag{4.4}$$

$$= d + \sum_{i=0}^{d-1} m_i + \sum_{i=0}^{d-1} n_i.$$

Since each of the systems are disjoint, the total number of indeterminates must be equal to the sum of the number of indeterminates in each of the components for both $x(u)$ and $y(u)$, yielding

$$m + 1 = \sum_{i=0}^{d-1} (m_i + 1) = d + \sum_{i=0}^{d-1} m_i,$$

from which

$$\sum_{i=0}^{d-1} m_i = m - d + 1 \tag{4.5}$$

is obtained. Similarly,

$$\sum_{i=0}^{d-1} n_i = n - d + 1 \tag{4.6}$$

can be derived. Substituting (4.5) and (4.6) into (4.4) yields

$$\mu_B(PM(m, n, d); G) = d + (m - d + 1) + (n - d + 1) = m + n - d + 2. \quad \blacksquare$$

The above proof did not depend on the index of the starting coefficient of the decimation, hence the multiplicative complexity only depends on the degrees of the two polynomials and the decimating factor. The choice of starting point will affect the types of systems that result from the partitioning. For example, if $PM(4, 4, 2)$ had a starting point of 1, then it decomposes into $PM(1, 2)$ and $PM(2, 1)$ instead of $PM(2, 2)$ and $PM(1, 1)$, but in both cases the number of required multiplications is 8.

The method of decomposition suggested by Theorem 4.4 can be useful in the implementation of algorithms for $PM(m, n, d)$. Obviously, an algorithm for $PM(m, n)$ could be used, requiring only $d-1$ more multiplications than is minimal, but one major benefit of recognizing $PM(m, n, d)$ as being composed of smaller polynomial multiplications is that each of these algorithms can choose from the same set of constants for the interpolation points in the Lagrange interpolation, resulting in a significant reduction in the number of additions required for this implementation. Minimal polynomial multiplication algorithms for even moderate values of m and n require large constants. For example if $G = Q$, the field of rational numbers, then an attempt at ordering constants from simplest to most complicated would be

$$0, \infty, 1, -1, 2, -2, 1/2, -1/2, 3, -3, 1/3, -1/3, 2/3, \cdots,$$

where beyond -1 the sequence is composed of sets of four elements of the form $\{a/b, -a/b, b/a, -b/a\}$ for relatively prime integers a and b.

It may also be possible to reduce the number of additions in the proposed algorithms for decimated polynomial multiplication by noticing that for large d many of the component polynomial multiplications have identical m and n and the polynomial results are summed to form the desired coefficients. Winograd [47] suggests that when a system has this structure and identical algorithms are used to compute the similar subsystems, then the reconstruction portion of the Chinese remainder theorem for each of the components can be combined by summing the CRT residues from these components and then doing only a single reconstruction. If the vector of residues for the ith component is r_i, the reconstruction matrix is C, and there are a total of k components, then this suggestion amounts to replacing the computation of $Cr_1 + Cr_2 + \cdots + Cr_k$ with $C(r_1 + r_2 + \cdots + r_k)$ by simply applying the distributive law.

4.5.2. Other Output Restrictions

Suppose that only the initial few coefficients of a polynomial product are desired and that the remainder may be ignored. The results of §3.2 can be applied here, since this truncation of the outputs is equivalent to multiplication of the polynomials $(\bmod \, u^{r+1})$, where r is the coefficient of largest degree that is desired. Following the reasoning of Theorem 4.1, one or both input polynomials has a rational root at infinity. Since u is an irreducible polynomial over G, and u^{r+1} is thus a power of an irreducible polynomial, then Theorem 3.2 states that for $T(u) = x(u)y(u) = \sum_{i=0}^{m+n} z_i u^i$,

$$\mu_B(z_0, z_1, \ldots, z_r; G) = \begin{cases} 2r+1, & r \le m, r \le n \\ n+r+1, & r \le m, r > n \\ m+r+1, & r > m, r \le n \\ m+n+1, & r > m, r > n \end{cases}$$

Theorem 4.5. *If* $T(u) = x(u)y(u) = \sum_{i=0}^{m+n} z_i u^i$, $x(u)$ *and* $y(u)$ *are polynomials with indeterminate coefficients of degree* m *and* n *respectively, and* $0 \le r \le m+n$, *then* $\mu_B(z_0, z_1, \ldots, z_r; G) = \min(m, r) + \min(n, r) + 1$.

Proof. This theorem is a summary of the results presented in the previous paragraph. ∎

This is a discouraging result, since it implies that if only half of the outputs are desired then frequently the same number of m/d steps are necessary as when the complete set of output coefficients is computed.

Suppose instead that the final few coefficients are desired, that is the set $\{z_r, z_{r+1}, \ldots, z_{m+n}\}$. The outputs of the system can be reversed to obtain a system equivalent to a polynomial product $(\mathrm{mod}\, u^{m+n-r+1})$. An analysis similar to that of Theorem 4.5 produces $\mu_B(z_r, z_{r+1}, \ldots, z_{m+n}; G) = 2m+2n-\max(m,r)-\max(n,r)+1$.

4.6. Summary of Chapter 4

In this chapter various input constraints and output restrictions have been imposed on systems of polynomial multiplication to determine their effects on the multiplicative complexity. It was shown that for the constrained polynomial of degree equal to or less than the unconstrained polynomial the multiplicative complexity is less than a general system of polynomial multiplication only if a polynomial in $G[u]$ divides the constrained polynomial (or the sum of the constrained polynomial and another polynomial in $G[u]$). When this is the case the multiplicative complexity is reduced by the degree of the polynomial factor from $G[u]$.

When the degree of the constrained polynomial is greater than the degree of the unconstrained polynomial, an upper bound on the multiplicative complexity can be determined by finding a representation of the constrained polynomial as a sum of products of polynomials in $G[u]$ with polynomials having indeterminate coefficients. The result is a direct sum of systems of polynomial products and the upper bound on the multiplicative complexity is simply the sum of the complexities of the components. Clearly, there is a "minimal" representation in the sense that no other decomposition of this sort uses fewer m/d steps. This minimal representation is not necessarily unique.

The results on constrained inputs were applied to the computation of the products of symmetric or antisymmetric polynomials with general polynomials. The multiplicative complexity of these systems is bounded above by the complexity of a system of general polynomial multiplication and can be less than this bound by only one or two depending on whether the degrees of the polynomials are even or odd. When both of the polynomials are either symmetric or antisymmetric, then the number of required m/d steps is between $\dfrac{m+n-2}{2}$ and $\dfrac{m+n+2}{2}$, or almost exactly half that required by general polynomial multiplication.

Two types of restrictions on the outputs of a polynomial product were considered. Decimating the output by some integer factor d results in a reduction of the multiplicative complexity by $d-1$ over the computation of the complete set of outputs. Truncating the set of outputs to the first r is equivalent to computing a polynomial product modulo u^{r+1} and the multiplicative complexity is equal to $\min(m,r)+\min(n,r)+1$. A similar result is obtained for computing the higher order coefficients only.

Several remaining questions must be addressed to complete a study of constraints on the inputs or restrictions on the outputs of polynomial products. Is the upper bound on the number of m/d steps obtained when the degree of the constrained polynomial is greater than the degree of the unconstrained polynomial equal to the lower bound? Are there general results for when both polynomials are constrained? An upper bound can obviously be established by removing factors with coefficients in G from each polynomial, multiplying the remaining polynomials, and finally multiplying by the removed factors. Is this also a lower bound?

Can anything be stated about arbitrary restrictions on the output? What if linear combinations (over G) of the outputs are desired rather than the coefficients of the polynomial product themselves? Finally, how do these results apply to polynomial products modulo other polynomials?

CHAPTER 5

Multiplicative Complexity of Discrete Fourier Transform

In this chapter the multiplicative complexity of the discrete Fourier transform (DFT) is analyzed. The next several sections define the DFT and then show how the complexity of the DFT is determined when the number of inputs is prime, a power of an odd prime, a power of two, and finally for any positive integer.

5.1. The Discrete Fourier Transform

Mathematically, for a vector whose entries are $x_0, x_1, \ldots, x_{N-1}$, the samples of the DFT are

$$X_k = \sum_{n=0}^{N-1} x_n e^{-j2\pi nk/N}, \quad k = 0, 1, \ldots, N-1. \tag{5.1}$$

Using Euler's relation, (5.1) can be expressed as

$$X_k = \sum_{n=0}^{N-1} x_n(\cos 2\pi nk/N - j\sin 2\pi nk/N), \quad k = 0, 1, \ldots, N-1,$$

emphasizing the trigonometric nature of the transformation. The DFT is commonly used to obtain a representation of the *frequency content* of a signal.

The DFT, or a formula closely resembling it, was apparently originally discovered by Euler [21] in 1750 in his investigations of sound propagation in elastic media. Shortly after Euler's discovery, formulas similar to the DFT were independently discovered by D. Bernoulli, Clairaut, and Lagrange. Their applications included the analysis of the motion of a vibrating string, sound propagation, and modeling the apparent motion of the sun relative to the earth.

Several decades passed before further effort was expended on trigonometric series. New research in this area was discouraged because it was not known whether these trigonometric series converged when the number of terms was infinite. In addition to the technical arguments against the use of trigonometric series, the late 18th century was also a bad time to be associated with scientific or mathematical research because of the political situation, particularly in France during the French Revolution.

Fourier began investigating the use of trigonometric series in developing an analytic theory of heat, possibly as early as 1802. His ideas were rejected by the French Academy of Sciences when he presented them in 1807, because he had no proof for the convergence of the series. Fourier persisted and ultimately convinced his critics that arbitrary functions defined over finite intervals could be expressed as infinite trigonometric series. Dirichlet investigated the convergence properties of trigonometric series and showed that Fourier's claims were mathematically sound. Fourier's research of trigonometric series led to them being called *Fourier series* and their extension to the *Fourier transform*.

In 1805, Gauss discovered a method for computing the DFT that he claimed "greatly reduces the tediousness of mechanical calculation" [21]. Gauss had derived a technique for computing the DFT that is now known as the *fast Fourier transform (FFT)*, that was rediscovered by Cooley and Tukey in 1965. Gauss applied the DFT to the interpolation of orbits of asteroids, but later discovered a better analytical method for solving this interpolation problem and did not publish his notes on the FFT during his lifetime. The notes do appear in Gauss' collected work, but were never referenced by later researchers except for a footnote in a mathematical encyclopedia in 1904. In 1977, Goldstine [19] realized the significance of Gauss' discovery and pointed out the connection with the algorithm of Cooley and Tukey.

The algorithm known as the fast Fourier transform (FFT) is only efficient when the length of the sequence (vector) to be transformed is a composite number. The next section presents results of Winograd and Rader that show how efficient algorithms can be derived for computing the DFT of sequences whose length is a prime number. The later sections show how the theory of multiplicative complexity can be applied to the computation of the DFT for other sequence lengths.

5.2. Prime Lengths

When the length of the sequence to be transformed is prime, Rader [33] has shown that the DFT may be reformulated as a cyclic convolution of length one less than the DFT length plus some auxiliary additions. Winograd [45] has analyzed the multiplicative complexity of the prime-length DFT, using the conversion to cyclic convolution. In the following analysis all references to prime numbers are to odd primes. We begin by considering only real-valued input sequences.

5.2.1. Rader's Permutation

The DFT of prime length is simply (5.1) with $N = p$, a prime. Rader showed that this computation may be reorganized by first separating the terms in this set of sums for which either n or k is zero (and $e^{-j2\pi nk/p} = 1$), yielding

$$X_0 = \sum_{n=0}^{p-1} x_n$$

$$X_k = x_0 + \sum_{n=1}^{p-1} x_n e^{-j2\pi nk/p}, \quad k = 1, 2, \ldots, p-1.$$

(5.2)

Let $w_p = e^{-j2\pi/p}$ be a pth root of unity. By definition, $w_p^p = 1$, and therefore $w_p^{nk} = w_p^{n'k'}$ if and only if $nk \equiv n'k' \pmod{p}$. The set of reduced residues (all residues excluding zero) modulo a prime p forms a group under multiplication. This group has $\phi(p-1)$ distinct primitive roots [26], where $\phi(\cdot)$ is Euler's totient function, and is isomorphic to the additive group Z_{p-1} of integers modulo $p-1$.

The existence of a primitive root g in this multiplicative group implies that all elements of the group may be expressed as g^i, $i = 0, 1, \ldots, p-2$. Rader suggests replacing n and k in (5.2) with $\langle g^n \rangle_p$ and $\langle g^k \rangle_p$ where the notation $\langle \cdot \rangle_p$ is a shorthand for reduction to the principal residue modulo p (i.e., $\langle i \rangle_p \equiv i \pmod p$ and $0 \le \langle i \rangle_p < p$). This replacement yields

$$X_0 = \sum_{n=0}^{p-1} x_{\langle g^n \rangle_p}$$

$$X_{\langle g^k \rangle_p} = x_0 + \sum_{n=0}^{p-2} x_{\langle g^n \rangle_p} w_p^{g^n g^k}, \quad k = 0, 1, \ldots, p-2,$$

but $g^n g^k = g^{n+k}$ and

$$X_{\langle g^k \rangle_p} = x_0 + \sum_{n=0}^{p-2} x_{\langle g^n \rangle_p} w_p^{g^{n+k}}, \quad k = 0, 1, \ldots, p-2. \tag{5.3}$$

The summations in (5.3) can be recognized as the circular correlation of the sequences $x_{\langle g^n \rangle_p}$ and $w_p^{g^n}$ with x_0 added to each output.

This formulation is called *Rader's permutation* because the input and output vectors are permuted according to the powers of the primitive root.

5.2.2. Multiplicative Complexity

In 1976, Winograd [42, 44] applied the theory of multiplicative complexity to the DFT of odd prime length to obtain efficient algorithms for the shortest lengths. He then used the tensor product ideas presented in the discussion of Definition 2.10 to merge algorithms for relatively prime lengths, resulting in new efficient algorithms for long composite lengths. These tensor product algorithms comprise what is now known as the Winograd Fourier transform algorithm. Winograd [45] subsequently carefully analyzed the multiplicative complexity of the discrete Fourier transform of prime length, hereafter referred to as DFT(p).

Winograd rediscovered Rader's permutation, but expressed the resulting equations in a slightly different form that emphasizes the direct sum structure of the transformation. Winograd replaced (5.2) with

$$X_0 = \sum_{n=0}^{p-1} x_n$$

$$X_{\langle g^k \rangle_p} = X_0 + \sum_{n=0}^{p-2} x_{\langle g^n \rangle_p} (w_p^{g^{n+k}} - 1), \quad k = 0, 1, \ldots, p-2.$$

(5.4)

Let P_I be the $(p-1) \times (p-1)$ input permutation matrix with $P_{I_{i, \langle g^{i-1} \rangle_p}} = 1$ for $i = 1, 2, \ldots, p-2$, and $P_{I_{i,j}} = 0$ otherwise. Let P_O be the $(p-1) \times (p-1)$ output permutation matrix with $P_{O_{\langle g^{p-2-i} \rangle_p, i}} = 1$ for $i = 1, 2, \ldots, p-2$, and $P_{O_{i,j}} = 0$ otherwise (this is the reverse order of (5.4), resulting in cyclic convolution rather than cyclic correlation). The transform of (5.4) can be represented in matrix notation as

$$\begin{bmatrix} X_0 \\ X_1 \\ \vdots \\ X_{p-1} \end{bmatrix} = \begin{bmatrix} 1 & 0 & \cdots & 0 \\ 1 & & & \\ \vdots & & P_O & \\ 1 & & & \end{bmatrix} \begin{bmatrix} 1 & 0 & \cdots & 0 \\ 0 & & & \\ \vdots & & \overline{W}_p & \\ 0 & & & \end{bmatrix} \begin{bmatrix} 1 & 1 & \cdots & 1 \\ 0 & & & \\ \vdots & & P_I & \\ 0 & & & \end{bmatrix} \begin{bmatrix} x_0 \\ x_1 \\ \vdots \\ x_{p-1} \end{bmatrix},$$

(5.5)

where

$$\overline{W}_p = \begin{bmatrix} \overline{w}_0 & \overline{w}_1 & \cdots & \overline{w}_{p-2} \\ \overline{w}_{p-2} & \overline{w}_0 & \cdots & \overline{w}_{p-3} \\ \vdots & \vdots & \ddots & \vdots \\ \overline{w}_1 & \overline{w}_2 & \cdots & \overline{w}_0 \end{bmatrix}$$

and $\overline{w}_i = w_p^{g^i} - 1$, $i = 0, 1, \ldots, p-2$.

Based on the definitions of the input and output permutation matrices, they are clearly transposes of one another. Therefore the input and output transformations in (5.5) are also transposes, a property that in some sense generalizes to all DFT lengths.

By Corollary 3.1, the number of m/d steps necessary and sufficient to implement cyclic convolution of length N (over the rational numbers) is $2N - \tau(N)$, where $\tau(N)$ is the number of positive divisors of N. This result does not apply directly here since the entries of the matrix $\overline{W}_p$ are not independent over Q. Theorem 3.9 must be applied in the analysis of this system.

The Chinese remainder theorem describes the decomposition of the cyclic convolution of length N into polynomial products modulo each of the irreducible factors (cyclotomic polynomials) dividing $u^N - 1$. Since $p-1$ is even, $u^{p-1} - 1 = (u^{(p-1)/2} - 1)(u^{(p+1)/2} - 1)$ is a partial factorization of the modulus polynomial. If g is a primitive root modulo p, then $g^{(p-1)/2} \equiv -1 \pmod{p}$ and thus

$$\sum_{i=0}^{p-2} \overline{w}_i u^i \equiv \sum_{i=0}^{p-2} (w_p^{g^i} - 1) u^i$$

$$\equiv \sum_{i=0}^{(p-3)/2} (w_p^{g^i} - 1 + w_p^{-g^i} - 1) u^i \tag{5.6}$$

$$\equiv 2 \sum_{i=0}^{(p-3)/2} (\cos \frac{2\pi g^i}{p} - 1) u^i \quad (\mathrm{mod}\, u^{(p-1)/2} - 1)$$

and

$$\sum_{i=0}^{p-2} \overline{w}_i u^i \equiv 2j \sum_{i=0}^{(p-3)/2} \sin \frac{2\pi g^i}{p} u^i \quad (\mathrm{mod}\, u^{(p-1)/2} + 1). \tag{5.7}$$

This decomposition has separated the system in such a way that half is purely real and the other half purely imaginary. Further decomposition of these factors into cyclotomic polynomials and subsequent reduction of the intermediate residues in (5.6) and (5.7) will still yield residues that are purely real or purely imaginary.

The preceding remarks show that the complex entries in $\overline{W}_p$ can be ignored in the analysis of the multiplicative complexity since the residue reductions yield products of polynomials with purely real coefficients (if the multiplication by j is commuted outside the imaginary polynomial products). The following theorem analyzes DFT(p) using Theorem 3.9 by determining the row rank of $\overline{W}_p$ over Q. The original proof is due to Winograd.

Theorem 5.1. [45] *For an input vector of p indeterminates, p an odd prime number,* $\mu_B(\mathrm{DFT}(p); Q) = 2p - \tau(p-1) - 3.$

Proof. The system DFT(p) is equivalent to $\bigcup_{j=1}^{\tau(p-1)} C(K_j; f_j, l_j)$ where $K_j = G[u]/\langle C_{d(j)}(u) \rangle$, $d(j)$ is the jth element of $\{d \ni d \mid (p-1)\}$, the entries of f_j are the coefficients of $\overline{w}_0 + \sum_{i=1}^{p-2} \overline{w}_{p-1-i} u^i \,(\mathrm{mod}\ C_{d(j)}(u))$, and the entries of l_j are the coefficients of $\sum_{i=0}^{p-2} x_{\langle g^i \rangle_p} u^i \,(\mathrm{mod}\, C_{d(j)}(u))$, $j = 1, 2, \ldots, \tau(p-1)$. This system may also be expressed as $\bigcup_{d \mid (p-1)} C(C_d; f_d, l_d)$ or $\bigcup_{j=1}^{\tau(p-1)} A_j(f_j) M^{(j)} y$.

The field F is $Q(w_p)$, and $[Q(w_p): Q] = p-1$, since the polynomial generating F is $C_p(u)$ and $\deg C_p = \phi(p) = p-1$. Exactly one basis element of $Q(w_p)$ can always be chosen to be in Q, thus $\dim L_Q(\bigcup_{j=1}^{\tau(p-1)} r(f_l)) = p-2$. Therefore, each of the f_j must have row rank $\phi(d(j))$, except one that is $\phi(d(j))-1$. All integers are divisible by one, thus $d(j) = 1$ for some j. This j always has $K_j = G[u]/\langle u-1 \rangle$, since $\sum_{i=1}^{p-1} w_p^i = -1$.

The product modulo $u-1$ in the union is a rational multiplication and may be eliminated from the complexity analysis. The remaining system is qrc reduced and $\dim L_Q(\underset{l\in L}{\cup} r(f_l)) = \sum_{l\in L} s_l \geq \sum_{l\in L} (s_l-1)$ for any subset L of $\{1, 2, \ldots, \tau(p-1)\}$ that does not include j with $d(j) = 1$. The cardinality of the remaining vector of indeterminates is $p-2$. Therefore, all the conditions of Theorem 3.9 are satisfied and

$$\mu_B(\text{DFT}(p); Q) \geq p-2+ \sum_{\substack{d|(p-1) \\ d\neq 1}} (\phi(d)-1)$$

$$= p-2+1-\phi(1)+ \left[\sum_{d|(p-1)} \phi(d)\right] -(\sum_{d|(p-1)} 1)$$

$$= p-2+(p-1)-\tau(p-1)$$

$$= 2p-\tau(p-1)-3.$$

We conclude that $\mu_B(\text{DFT}(p); Q) = 2p-\tau(p-1)-3$, since the algorithm used in Theorem 3.3 achieves the bound. ■

5.3. Powers of Prime Lengths

When the sequence length N is a power of an odd prime number, the part of the DFT computation that relates outputs with indexes relatively prime to N to the set of inputs with the same indexes is equivalent to a cyclic convolution [24]. The multiplicative complexity of this part of the computation of the length-p^r DFT has been analyzed by Winograd [45]. Recently, Auslander et al. [4,5] decomposed the length-p^r DFT into a union of polynomial products modulo irreducible polynomials, from which the multiplicative complexity of the length-p^r DFT may be deduced by Theorem 3.9.

Rader's permutation can be extended to DFT lengths that are powers of odd primes. The existence of primitive roots in the multiplicative group of the reduced residue system modulo p^r can be used, as in (5.3), to convert part of the computation of DFT(p^r) into cyclic convolution. This cyclic convolution relates the set of outputs with indexes relatively prime to p^r to the set of inputs with indexes relatively prime to p^r. This is a $(p^r-p^{r-1})\times(p^r-p^{r-1})$ square submatrix of the DFT(p^r) matrix called the core of the DFT or, CFT(p^r) [47].

Primitive roots exist only for the reduced residue systems modulo p^r, $2p^r$, 2, and 4, for p an odd prime and r a positive integer [26]. CFT(N) is a cyclic convolution only for DFTs of these lengths. For other lengths, with $N = \prod_{i=1}^{m} p_i^{e_i}$, CFT($N$) is equivalent to the direct product $\overset{m}{\underset{i=1}{\otimes}} \text{CFT}(p_i^{e_i})$.

Let $\overline{w}_{i;j} = w_{p^j}^{g^i} - 1$, $j = 1, 2, \ldots, r$, $i = 0, 1, \ldots, p^j - p^{j-1} - 1$. The submatrix $\text{CFT}(p^r)$ is a cyclic convolution matrix (circulant),

$$\text{CFT}(p^r) = \begin{bmatrix} \overline{w}_{0;r} & \overline{w}_{1;r} & \cdots & \overline{w}_{p^r-p^{r-1}-1;r} \\ \overline{w}_{p^r-p^{r-1}-1;r} & \overline{w}_{0;r} & \cdots & \overline{w}_{p^r-p^{r-1}-2;r} \\ \vdots & \vdots & \ddots & \vdots \\ \overline{w}_{1;r} & \overline{w}_{2;r} & \cdots & \overline{w}_{0;r} \end{bmatrix}.$$

If every p^{th} output, starting from zero, of $\text{DFT}(p^r)$ is computed, the resulting system is equivalent to $\text{DFT}(p^{r-1})$. From this subsystem, we can extract $\text{CFT}(p^{r-1})$, and recursively come up with smaller subsystems until $\text{CFT}(p)$ is reached.

The problem encountered in subdividing the matrix $\text{DFT}(p^r)$ into CFT submatrices is that this representation of the system does not appear to be a union of systems of polynomial multiplication. As an example, a permutation of the inputs and outputs of DFT(9) (using the primitive root $g = 2$ of 9) yields

$$\begin{bmatrix} X_0 \\ X_3 \\ X_6 \\ X_1 \\ X_5 \\ X_7 \\ X_8 \\ X_4 \\ X_2 \end{bmatrix} = \begin{bmatrix} 0 & 0 & 0 & 0 & 0 & 0 & 0 & 0 & 0 \\ 0 & 0 & 0 & 3 & 6 & 3 & 6 & 3 & 6 \\ 0 & 0 & 0 & 6 & 3 & 6 & 3 & 6 & 3 \\ 0 & 3 & 6 & 1 & 2 & 4 & 8 & 7 & 5 \\ 0 & 6 & 3 & 5 & 1 & 2 & 4 & 8 & 7 \\ 0 & 3 & 6 & 7 & 5 & 1 & 2 & 4 & 8 \\ 0 & 6 & 3 & 8 & 7 & 5 & 1 & 2 & 4 \\ 0 & 3 & 6 & 4 & 8 & 7 & 5 & 1 & 2 \\ 0 & 6 & 3 & 2 & 4 & 8 & 7 & 5 & 1 \end{bmatrix} \begin{bmatrix} x_0 \\ x_3 \\ x_6 \\ x_1 \\ x_2 \\ x_4 \\ x_8 \\ x_7 \\ x_5 \end{bmatrix}$$

where the entries of the matrix are the exponents of w_9. The lower right 6×6 submatrix is CFT(9) and six occurrences of CFT(3) are apparent, three in the two rows above CFT(9) and three in the two columns to the left of CFT(9). All other entries of the DFT(9) matrix are equal to one and do not influence the multiplicative complexity.

The three CFT(3) blocks in the same two rows may be reduced to a single CFT(3) block by using the distributive law,

$$\text{CFT(3)} \begin{bmatrix} x_1 \\ x_2 \end{bmatrix} + \text{CFT(3)} \begin{bmatrix} x_4 \\ x_8 \end{bmatrix} + \text{CFT(3)} \begin{bmatrix} x_7 \\ x_5 \end{bmatrix} = \text{CFT(3)} \begin{bmatrix} x_1+x_4+x_7 \\ x_2+x_8+x_5 \end{bmatrix}. \tag{5.8}$$

The three CFT(3) blocks in the same two columns may also be reduced to a single

CFT(3) block by computing the product $\text{CFT(3)} \begin{bmatrix} x_3 \\ x_6 \end{bmatrix}$ once and then adding the result

to the other part of the computation of $[X_1 \, X_5 \, X_7 \, X_8 \, X_4 \, X_2]^T$.

The difficulty in analyzing this system is that the two remaining CFT(3) blocks share inputs in one case and outputs in the other case with the CFT(9) block. The system DFT(9) does not appear to be a union of smaller systems. The following analysis will show that DFT(9) does decompose into a union of polynomial products, and in general DFT(p^r) always decomposes in a similar manner.

Let $R(u) = x_1 + x_2 u + x_4 u^2 + x_8 u^3 + x_7 u^4 + x_5 u^5$ and let $S(u) = w_9^1 + w_9^5 u + w_9^7 u^2 + w_9^8 u^3 + w_9^4 u^4 + w_9^2 u^5$, so that the cyclic convolution implicit in CFT(9) is the polynomial product

$$R(u)S(u) \ (\text{mod } u^6-1).$$

The sums computed in (5.8) are the same terms that result from reducing $R(u)$ modulo u^2-1. A similar reduction of $S(u)$ yields

$$S(u) \equiv (w_9^1 + w_9^7 + w_9^4) + (w_9^5 + w_9^8 + w_9^2)u$$
$$\equiv (w_9^0 + w_9^3 + w_9^6)(w_9^1 + w_9^2 u)$$
$$\equiv (w_3^0 + w_3^1 + w_3^2)(w_9^1 + w_9^2 u) \ (\text{mod } u^2-1).$$

The term $w_3^0 + w_3^1 + w_3^2$ is the polynomial u^2+u+1 evaluated at w_3, and must be zero since u^2+u+1 is the minimal (cyclotomic) polynomial of w_3. Thus $R(u)S(u) \equiv 0 \ (\text{mod } u^2-1)$ and CFT(9) reduces to the system $R(u)S(u) \ (\text{mod } u^4+u^2+1)$, or $MPM(u^4+u^2+1)$.

Before completing the analysis of $\mu_B(\text{DFT(9)})$, it is useful to see how the decomposition of DFT(9) generalizes for DFT(p^r). Consider first the subsystem that computes all outputs with indexes divisible by p. This subsystem is

$$X_{pk_1} = \sum_{n=0}^{p^r-1} x_n w_{p^r}^{npk_1}$$
$$= \sum_{n_2=0}^{p^{r-1}-1} \left[\sum_{n_1=0}^{p-1} x_{p^{r-1}n_1+n_2} \right] w_{p^{r-1}}^{n_2 k_1}, \quad k_1 = 0, 1, \ldots, p^{r-1}-1 \tag{5.9}$$

and is obviously the length-p^{r-1} DFT of the sequence $\tilde{x}_{n_2} = \sum_{n_1=0}^{p-1} x_{p^{r-1}n_1+n_2}$,

$$n_2 = 0, 1, \ldots, p^{r-1} - 1.$$

The remaining outputs are the set whose indexes are relatively prime to the sequence length, and thus the extension of Rader's permutation can be applied. Considering all inputs simultaneously,

$$X_{\langle g^k \rangle_{p^r}} = \sum_{n=0}^{p^r-1} x_n w_{p^r}^{ng^k}$$

$$= x_0 + \sum_{i=1}^{r} \left[\sum_{n=0}^{p^i - p^{i-1} - 1} x_{\langle p^{r-i} g^n \rangle_{p^r}} w_{p^i}^{g^{n+k}} \right], \quad k = 0, 1, \ldots, p^r - p^{r-1} - 1, \tag{5.10}$$

where g is a primitive root for all powers of p (existence of such a root is shown in [26]). This subset of the outputs is related to the inputs through the sum of r cyclic convolutions of lengths $p^i - p^{i-1}$, $i = 1, 2, \ldots, r$.

The factorization

$$u^N - 1 = (u-1)(\sum_{k=0}^{N-1} u^k)$$

holds for all positive integers N. For $N = p$,

$$u^p - 1 = (u-1)(\sum_{k=0}^{p-1} u^k). \tag{5.11}$$

A factorization of $u^{p^i - p^{i-1}} - 1$ may be obtained by replacing u with $u^{p^{i-1} - p^{i-2}}$ in (5.11), yielding

$$u^{p^i - p^{i-1}} - 1 = (u^{p^{i-1} - p^{i-2}} - 1)(\sum_{k=0}^{p-1} u^{k(p^{i-1} - p^{i-2})}) \tag{5.12}$$

which holds for $i \geq 2$. The two factors in (5.12) can be easily resolved into irreducible factors using the definition of cyclotomic polynomials given in Appendix A,

$$u^{p^i - p^{i-1}} - 1 = \prod_{d \mid (p^i - p^{i-1})} C_d(u)$$

$$= \prod_{d \mid (p-1)} \left[\prod_{k=0}^{i-1} C_{p^k d}(u) \right], \tag{5.13}$$

where $C_d(u)$ is the dth cyclotomic polynomial. From (5.13) it follows that

$$u^{p^{i-1} - p^{i-2}} - 1 = \prod_{d \mid (p-1)} \left[\prod_{k=0}^{i-2} C_{p^k d}(u) \right].$$

and

$$\sum_{k=0}^{p-1} u^{k(p^{i-1}-p^{i-2})} = \frac{u^{p^i-p^{i-1}}-1}{u^{p^{i-1}-p^{i-2}}-1}$$

$$= \prod_{d|(p-1)} C_{p^{i-1}d}(u)$$

are the factorizations into irreducible polynomials.

One set of polynomials in (5.10) is

$$\sum_{n=0}^{p^i-p^{i-1}-1} w_{p^i}^{g^n} u^n, \quad i = 1, 2, \ldots, r. \tag{5.14}$$

Each of the polynomials in (5.14) can be reduced modulo $u^{p^{i-1}-p^{i-2}}-1$ to obtain

$$\sum_{n=0}^{p^i-p^{i-1}-1} w_{p^i}^{g^n} u^n \equiv \sum_{n=0}^{p^{i-1}-p^{i-2}-1} \left[\sum_{l=0}^{p-1} w_{p^i}^{g^{l(p^{i-1}-p^{i-2})+n}} \right] u^n \pmod{u^{p^{i-1}-p^{i-2}}-1}. \tag{5.15}$$

Niven and Zuckerman [26] show that $g^{p^{i-1}-p^{i-2}} = 1+p^{i-1}n_i$, where $p \nmid n_i$ and $i \geq 2$. Therefore the exponents of the roots of unity inside parentheses in (5.15) are

$$g^{l(p^{i-1}-p^{i-2})+n} = (1+p^{i-1}n_i)^l g^n$$

$$\equiv g^n \pmod{p^{i-1}}. \tag{5.16}$$

Each of the powers of w_{p^i} in (5.14) is distinct, implying that each of the powers of w_{p^i} in (5.15) is also distinct. Exactly p distinct residues modulo p^i exist that are congruent to g^n modulo p^{i-1}, as in (5.16). The sum over l in (5.15) contains p exponents each of which is congruent to g^n modulo p^{i-1} and distinct from the others and since only p distinct residues exist that satisfy this congruence, the entire set must be represented in this sum. Therefore the inner sum of (5.15) may be rewritten as

$$\sum_{l=0}^{p-1} w_{p^i}^{g^{l(p^{i-1}-p^{i-2})+n}} = \sum_{l=0}^{p-1} w_{p^i}^{lp^{i-1}+g^n}$$

$$= w_{p^i}^{g^n} \sum_{l=0}^{p-1} w_p^l \tag{5.17}$$

$$= 0,$$

since w_p is a root of the pth cyclotomic polynomial, $C_p(u) = \sum_{i=0}^{p-1} u^i$.

The polynomials in (5.14) each satisfy a congruence

$$\sum_{n=0}^{p^i-p^{i-1}-1} w_{p^i}^{g^n} u^n \equiv 0 \ (\text{mod} \ u^{p^{i-1}-p^{i-2}}-1)$$

for $i = 2, 3, \ldots, r$, and for $i = 1$, the congruence

$$\sum_{n=0}^{p-2} w_p^{g^n} u^n \equiv -1 \ (\text{mod} \ u-1)$$

is satisfied. In computing the length p^i-p^{i-1} cyclic convolutions $(i = 2, 3, \ldots, r)$ implicit in (5.10), it is only necessary to reduce the input polynomials modulo $\sum_{k=0}^{p-1} u^{k(p^{i-1}-p^{i-2})}$, multiply the pairs of residues modulo $\sum_{k=0}^{p-1} u^{k(p^{i-1}-p^{i-2})}$, and then reconstruct residues modulo $u^{p^i-p^{i-1}}-1$ using the Chinese remainder theorem.

Excluding the single rational multiplication necessary to obtain the residue modulo $u-1$, the sum of the degrees of the remaining polynomial moduli is

$$\begin{aligned}
\sum_i \deg P_i &= p-2+ \sum_{i=2}^{r} (p-1)(p^{i-1}-p^{i-2}) \\
&= p-2+p^r-p^{r-1}-p+1 \\
&= p^r-p^{r-1}-1.
\end{aligned} \tag{5.18}$$

The following lemma will help to relate this to $\dim L_Q(r(w_{p^r}^0), r(w_{p^r}^1), \ldots, r(w_{p^r}^{p^{r}-1}))$.

Lemma 5.1. *For any positive integer* N, $\dim L_Q(w_N^0, w_N^1, \ldots, w_N^{N-1})$ $= [Q(w_N):Q] = \phi(N)$, *where* $w_N = e^{-j2\pi/N}$ *and* $\phi(N)$ *is Euler's totient function.*

Proof. Let $C_N(u)$ be the Nth cyclotomic polynomial and thus the minimal polynomial of w_N. Any polynomial with rational coefficients and root w_N must be divisible by $C_N(u)$ over the ring $Q[u]$, by definition of the minimal polynomial.

The set of polynomials $\Lambda = \{C_N(u), uC_N(u), \ldots, u^{N-\phi(N)-1}C_N(u)\}$ will be shown to be a set of $N-\phi(N)$ independent polynomials (over Q) that span the subset $\{P(u) \in Q[u] \ni \deg P(u) < N \text{ and } P(w_N) = 0\}$. The first element $C_N(u)$ of Λ is a nonzero $\phi(N)^{th}$ degree polynomial and thus provides an initial basis element for the space spanned by the elements of Λ. Each successive element of Λ is of higher degree than all previous elements and thus cannot be expressed as a linear combination of these previous elements over Q. Therefore if each element of Λ is considered in the order given, then the linear independence of all elements of Λ over Q is proved by induction.

Each of the polynomials in Λ has root w_N, yielding the equations

$$w_N^i C_N(w_N) = 0, \quad i = 0, 1, \ldots, N - \phi(N) - 1, \tag{5.19}$$

that constitute a set of $N - \phi(N)$ independent rational linear relationships between the w_N^k, $k = 0, 1, \ldots, N-1$. It remains to be shown that all other linear relationships (over Q) between the Nth roots of unity are dependent on this set.

Products of $C_N(u)$ and other polynomials in $Q[u]$ of degree less than $N - \phi(N)$ are clearly rational linear combinations of elements of Λ. Likewise, products of $C_N(u)$ and polynomials in $Q[u]$ of degree $N - \phi(N)$ or greater may be reduced modulo $u^N - 1$ (since $u = w_N$ is the only point at which the product is evaluated), but since $C_N(u)$ divides $u^N - 1$, the residue is the difference of two multiples of $C_N(u)$, and must be divisible by $C_N(u)$ itself. Thus no linear combination of the powers of w_N is equal to zero that is not dependent on the set of $N - \phi(N)$ independent equations in (5.19).

Therefore $\dim L_Q(w_N^0, w_N^1, \ldots, w_N^{N-1})$ must be equal to the total number of elements N less the $N - \phi(N)$ dependent entries, or

$$\dim L_Q(w_N^0, w_N^1, \ldots, w_N^{N-1}) = N - (N - \phi(N))$$

$$= \phi(N). \quad \blacksquare$$

Corollary 5.1. $\dim L_Q(r(w_N^0), r(w_N^1), \ldots, r(w_N^{N-1})) = \phi(N) - 1$.

Proof. The result of Lemma 5.1 gives the dimension of the linear span of this set, not taking into account the effect of the mapping r. The element $w_N^0 = 1$ is always rational, thus the set $\{w_N^i\}$ includes the rational numbers as a subset. Elements of this subset are mapped to zero by r, reducing the dimension of the linear span by one. The mapping r affects no other dimension of the space spanned by the set. Therefore the dimension of the linear span under the mapping r is exactly one less than the dimension before the mapping r is applied. $\blacksquare$

Corollary 5.1 holds for all values of N, but the conversion of the DFT matrix with its complex entries into real-valued submatrices as in (5.6) and (5.7) requires evaluation of the dimension over $Q(j)$, since any rational linear relations between the real and imaginary components will affect the total row rank as defined in Theorem 3.9. In the proof of Theorem A.1 in Appendix A it is shown that $[Q(w_N):Q(w_M)] = \dfrac{\phi([N,M])}{\phi(M)}$ where $[N, M] = \dfrac{NM}{(N,M)}$ is the least common multiple of N and M. It is also shown that $\phi(N)\phi(M) = \phi([N,M])\phi((N,M))$, yielding $[Q(w_N):Q(w_M)] = \dfrac{\phi(N)}{\phi((N,M))}$. Since j is the primitive 4th root of unity, $[Q(w_N):Q(j)] = \dfrac{\phi(N)}{\phi((N,4))}$.

Corollary 5.2. $\dim L_Q(r(\text{Re}(w_N^0)), r(\text{Re}(w_N^1)), \ldots, r(\text{Re}(w_N^{N-1})))$,

$$r(\text{Im}(w_N^0)), r(\text{Im}(w_N^1)), \ldots, r(\text{Im}(w_N^{N-1}))) = \frac{\phi(N)}{\phi((N,4))} - 1.$$

Proof. This result follows from $[Q(w_N):Q(j)] = \dfrac{\phi(N)}{\phi((N,4))}$ and the fact that the rationals constitute a one-dimensional subspace of the linear span that is mapped to zero by r as explained in the proof of Corollary 5.1. ∎

Corollary 5.2 shows that the total row rank of the polynomial product decomposition for DFT(N) is $\phi(N)-1$ when N is not a multiple of four, and is $\frac{1}{2}\phi(N)-1$ when N is a multiple of four. In considering DFT(p^r), the row rank is $\phi(p^r)-1 = p^r-p^{r-1}-1$, which is exactly the same as the sum of the degrees of the modulus polynomials determined in (5.18). Therefore this submatrix of DFT(p^r) decomposes into a union of polynomial products that is equivalent to the original submatrix.

The residue reductions in (5.17) that are zero correspond exactly to the column sums that were done to convert the remainder of DFT(p^r) into DFT(p^{r-1}). Therefore a nonsingular $p^r \times p^r$ G-matrix exists that separates DFT(p^r) into these two components as a union of systems. We are now prepared to evaluate the multiplicative complexity of DFT(p^r).

Theorem 5.2. *For an input vector of p^r indeterminates, p an odd prime number and r a positive integer,* $\mu_B(\text{DFT}(p^r); Q) = 2p^r - r - 2 - \dfrac{r^2+r}{2}\tau(p-1).$

Proof. The system DFT(p^r) has been shown to be equivalent to the union of the system DFT(p^{r-1}), and a system that is itself the union of several polynomial products modulo irreducible polynomials. The part of DFT(p^r) equivalent to DFT(p^{r-1}) can then be reduced to a union of DFT(p^{r-2}) and some polynomial products modulo the same set of irreducible polynomials previously mentioned, excluding the polynomial of largest degree. This decomposition can be continued recursively until the entire computation has been broken down into a union of polynomial products modulo the original set of irreducible polynomials. Each polynomial product with the same modulus polynomial also shares a polynomial multiplicand (i.e., the polynomial whose coefficients are rational linear combinations of the powers of w_{p^r}). Therefore the system DFT(p^r) is equivalent to the system

$$\bigcup_{j=1}^{r} \left[\bigcup_{i=0}^{j-1} \left[\bigcup_{\substack{d|p-1 \\ (i,d) \neq (0,1)}} C(C_{p^i d}(u); f_{i,d}, l_{j,i,d}) \right] \right]$$

where the entries of $f_{i,d}$ are the coefficients of

$$\sum_{n=0}^{p^{i+1}-p^i-1} w_{p^{i+1}}^{g^n} u^n \pmod{C_{p^i d}(u)}, \quad i=0,1,\ldots,r-1, \ d|(p-1), \ (i,d) \neq (0,1),$$

the entries of $l_{j,i,d}$ are the coefficients of

$$\sum_{n=0}^{p^{i+1}-p^i-1} \left[\sum_{l=0}^{p^{r-j}-1} x_{\langle lp^j + p^{j-1} g^{-n}\rangle_{p^r}} \right] u^n \pmod{C_{p^i d}(u)},$$

for $j=1,2,\ldots,r$, $i=0,1,\ldots,j-1$, $d|(p-1)$, $(i,d) \neq (0,1)$, and g is a primitive root of p^r. This system can be rewritten as

$$\bigcup_{i=0}^{r-1} \left[\bigcup_{\substack{d|(p-1) \\ (i,d)\neq(0,1)}} ((r-i)A_{i,d}(f_{i,d}))M^{(i,d)}y \right] \tag{5.20}$$

where $A_{i,d}(f_{i,d})$ is the regular representation of $f_{i,d}$, and $M^{(i,d)}y$ contains the linear combination of the inputs specified by $l_{j,i,d}$.

The system described by (5.20) is qrc reduced and $\dim L_Q(\bigcup_{l\in L} r(f_l)) = \sum_{l\in L} s_l \geq \sum_{l\in L}(s_l-1)$ for any subset L of (i,d) pairs, $i=0,1,\ldots,r-1$, $d|(p-1)$, $(i,d) \neq (0,1)$. The cardinality of y, excluding the modulo $u-1$ products, is equal to the sum of the degrees of all the polynomial moduli, or

$$m = \sum_{i=1}^{r} (p^i - p^{i-1} - 1)$$

$$= p^r - 1 - r.$$

Theorem 3.9 may now be applied to yield

$$\mu_B(DFT(p^r; Q)) \geq \sum_{\substack{d|(p-1)}} \sum_{\substack{i=0 \\ (i,d)\neq(0,1)}}^{r-1} (r-i)(s_{i,d}-1) + m$$

$$= 2m - \sum_{d|(p-1)} \sum_{i=0}^{r-1} (r-i) + r$$

$$= 2(p^r-1-r) + r - \sum_{i=1}^{r} i\tau(p-1)$$

$$= 2p^r - r - 2 - \frac{r^2+r}{2}\tau(p-1).$$

As for Theorem 5.1, the algorithm of Theorem 3.3 achieves the lower bound, and thus $\mu_B(DFT(p^r; Q)) = 2p^r - r - 2 - \frac{r^2+r}{2}\tau(p-1)$. ∎

One interesting aspect of the derivation of the multiplicative complexity of DFT(p^r) is that for each subset $\{X_{\langle p^{r-i}g^k\rangle_{p^r}}; k = 0, 1, \ldots, p^i - p^{i-1} - 1\}$, $i = 1, 2, \ldots, r$, the modulus polynomials for the polynomial products in that part of the union also constitute the set of irreducible factors of $u^{p^i - p^{i-1}} - 1$ excluding $u - 1$. If the rational multiplications associated with the factor $u - 1$ are included, then each of these subsystems is equivalent to a cyclic convolution. The system DFT(p^r) is thus equivalent to the system $\bigcup\limits_{i=1}^{r} MPM \left[\dfrac{u^{\phi(p^i)} - 1}{u - 1} \right]$.

The multiplicative complexity analysis of DFT(p^r) is not aided by this interpretation, but it provides an interesting perspective on the extension of Rader's permutation. There does not seem to be a computational advantage to be gained through this observation since ultimately the cyclic convolutions are broken down into their components in computing the system. Forming the coefficients of the cyclic convolutions also results in some rational multiplications that cannot be entirely commuted into the linear combinations of powers of w_{p^r} and thus increase the number of actual multiplications for systems in which rational multiplications are implemented as real multiplications.

5.4. Power-of-Two Lengths

As for the lengths already investigated, DFT(2^r) can be decomposed into a set of cyclic convolutions, and ultimately into a union of polynomial products modulo irreducible polynomials. It was pointed out in the previous section that no primitive root exists for the multiplicative group modulo 2^r ($r \geq 3$), so that cyclic convolution does not manifest itself in the same way as for DFT(p) and DFT(p^r).

DFT(2^r) decomposes into a system that computes the odd-indexed outputs, and a system equivalent to DFT(2^{r-1}) that computes the even-indexed outputs. This is identical to the decomposition of DFT(p^r) described in (5.9), and these even-indexed outputs are expressed as

$$X_{2k_1} = \sum_{n=0}^{2^r-1} x_n w_{2^r}^{2nk_1}, \quad k_1 = 0, 1, \ldots, 2^{r-1} - 1$$

$$= \sum_{n_2=0}^{2^{r-1}-1} (x_{n_2} + x_{2^{r-1}+n_2}) w_{2^{r-1}}^{n_2 k_1}, \quad k_1 = 0, 1, \ldots, 2^{r-1} - 1$$

which is obviously a length-2^{r-1} DFT.

Since the set of odd indexes modulo 2^r does not form a cyclic group, it is necessary to determine the algebraic structure of this reduced residue system. The multiplicative group (of the reduced residue system) modulo 2^r is known to be isomorphic to $Z_2 \times Z_{2^{r-2}}$, where Z_n denotes the additive group modulo n [26]. It is also well known that the order of 5 in the multiplicative group modulo 2^r is 2^{r-2} and therefore that the integers

$$\pm 5, \pm 5^2, \ldots, \pm 5^{2^{r-2}}$$

form a reduced residue system modulo 2^r for $r \geq 3$ [26].

Proceeding in the same way as for DFT(p^r), we obtain

$$X_{\langle 5^k(-1)^l\rangle_{2^r}} = \sum_{n=0}^{2^r-1} x_n w_{2^r}^{5^k(-1)^l n}$$

$$= x_0 - x_{2^{r-1}} + \sum_{i=2}^{r} \left[\sum_{n=0}^{2^{i-2}-1} \sum_{m=0}^{1} x_{\langle 2^{r-i}5^n(-1)^m\rangle_{2^r}} w_{2^i}^{5^{n+k}(-1)^{l+m}} \right], \qquad (5.21)$$

$$k = 0, 1, \ldots, 2^{r-2}, \; l = 0, 1.$$

The inner sums in (5.21) are each a direct product of a length-2 cyclic convolution with length-2^{i-2} cyclic convolutions, $i = 2, 3, \ldots, r$.

The fixed polynomial for the ith product is

$$\sum_{n=0}^{2^{i-2}-1} \sum_{m=0}^{1} w_{2^i}^{5^n(-1)^m} u^n v^m, \qquad (5.22)$$

the polynomial formed from the inputs is

$$\sum_{n=0}^{2^{i-2}-1} \sum_{m=0}^{1} x_{\langle 2^{r-i}5^{-n}(-1)^m\rangle_{2^r}} u^n v^m,$$

and the modulus polynomials are $u^{2^{i-2}}-1$ and v^2-1.

Again, as for DFT(p^r), (5.22) is divisible by one factor of the modulus polynomial $u^{2^{i-2}}-1$. This modulus polynomial can be factored into

$$u^{2^{i-2}}-1 = (u^{2^{i-3}}-1)(u^{2^{i-3}}+1).$$

Some properties of the group $Z_{2^{i-2}}$ need to be investigated to determine the residue of the expression (5.22) modulo $u^{2^{i-3}}-1$.

It is well known that products of two numbers of the form $4k+1$ are also of the form $4k+1$ since

$$(4k+1)(4l+1) = 4(4kl+k+l)+1. \qquad (5.23)$$

The number 5 is of the form $4k+1$. If $5^l \equiv 1 \pmod{4}$, then by (5.23) we have $5^{l+1} \equiv 1 \pmod{4}$, and by induction on l, beginning with $l = 1$, $5^l \equiv 1 \pmod{4}$ for any positive integer l.

The set of residues $\{5^l; l = 0, 1, \ldots, 2^{r-2}-1\}$ modulo 2^i are all distinct since 5 is of order 2^{i-2} modulo 2^i. Therefore only one square root of unity, other than unity itself, can be in this set; otherwise unity would be repeated. This square root must be $5^{2^{i-3}} \equiv 2^{i-1}+1 \pmod{2^i}$, since $(2^{i-1}+1)^2 = 2^{2i-2}+2^i+1 \equiv 1 \pmod{2^i}$ and $2^{i-1}+1 \equiv 1 \pmod 4$ (the other two square roots of unity are $2^{i+1}-1$ and 2^i-1, neither of which is congruent to 1 modulo 4 and thus cannot be congruent to a power of 5 modulo 2^i).

It is clear that $w_{2^i}^{2^{i-1}} = w_2 = -1$ for all $i \geq 1$. Reduction of (5.22) modulo $u^{2^{i-3}}-1$ for $3 \leq i \leq r$ yields

$$
\begin{aligned}
\sum_{n=0}^{2^{i-2}-1} \sum_{m=0}^{1} w_{2^i}^{5^n(-1)^m} u^n v^m &\equiv \sum_{n=0}^{2^{i-3}-1} \sum_{m=0}^{1} \left[w_{2^i}^{5^n(-1)^m} + w_{2^i}^{5^{n+2^{i-3}}(-1)^m} \right] u^n v^m \\
&\equiv \sum_{n=0}^{2^{i-3}-1} \sum_{m=0}^{1} \left[w_{2^i}^{5^n(-1)^m} + (w_{2^i}^{5^{2^{i-3}}})^{5^n(-1)^m} \right] u^n v^m \\
&\equiv \sum_{n=0}^{2^{i-3}-1} \sum_{m=0}^{1} \left[w_{2^i}^{5^n(-1)^m} + (w_{2^i}^{2^{i-1}+1})^{5^n(-1)^m} \right] u^n v^m \\
&\equiv \sum_{n=0}^{2^{i-3}-1} \sum_{m=0}^{1} \left[w_{2^i}^{5^n(-1)^m} + (-w_{2^i})^{5^n(-1)^m} \right] u^n v^m \\
&\equiv \sum_{n=0}^{2^{i-3}-1} \sum_{m=0}^{1} \left[w_{2^i}^{5^n(-1)^m} - w_{2^i}^{5^n(-1)^m} \right] u^n v^m
\end{aligned}
\tag{5.24}
$$

$$
\equiv 0 \pmod{u^{2^{i-3}}-1}.
$$

The length-2^{i-2} cyclic convolutions, $i = 3, 4, \ldots, r$, in (5.21) are equivalent to polynomial products modulo $u^{2^{i-3}}+1$. For $i = 2$, the polynomial in (5.22) is a constant, $w_{2^2} = -j$, and therefore this polynomial product requires no multiplications. The sum of the degrees of the remaining polynomial moduli is

$$
\begin{aligned}
\sum_i \deg P_i &= \sum_{i=3}^{r} 2^{i-3} \\
&= \sum_{i=0}^{r-3} 2^i \\
&= 2^{r-2}-1.
\end{aligned}
\tag{5.25}
$$

Reduction of (5.22) modulo $v-1$ (and $u^{2^{n-3}}+1$) yields

$$4 \sum_{n=0}^{2^{i-3}-1} \cos \frac{2\pi 5^n}{2^i} u^n, \tag{5.26}$$

and reduction of (5.22) modulo $v+1$ (and $u^{2^{n-3}}+1$) yields

$$-4j \sum_{n=0}^{2^{i-3}-1} \sin \frac{2\pi 5^n}{2^i} u^n. \tag{5.27}$$

From Corollary 5.2, the dimension of the space spanned by the cosines and sines in (5.26) and (5.27) is $2^{r-2}-1$. This is identical to the sum of the degrees of the modulus polynomials in u shown in (5.25). The total degree of all the polynomials, including both (5.26) and (5.27) is twice this, or $2^{r-1}-2$. In the following it will be shown that the two polynomial multiplications modulo $2^{i-3}+1$ can be rearranged so that the multiplying polynomials in (5.26) and (5.27) are identical.

We will show by induction that $5^{2^{i-4}} \equiv 3 \cdot 2^{i-2}+1 \pmod{2^i}$ for $i \geq 5$. Computing $5^{2^{5-4}} = 5^2 \equiv 25 \pmod{32}$ demonstrates the assertion for $i = 5$. Assume the assertion to be true for all $i \geq 5$, then $5^{2^{i-4}} \equiv 3 \cdot 2^{i-2}+1 \pmod{2^i}$ and therefore either

$$5^{2^{i-4}} \equiv 3 \cdot 2^{i-2}+1 \pmod{2^{i+1}} \quad \text{or}$$
$$5^{2^{i-4}} \equiv 2^i+3 \cdot 2^{i-2}+1 \pmod{2^{i+1}}.$$

For the first case,

$$5^{2^{(i+1)-4}} = \left[5^{2^{i-4}}\right]^2 \equiv (3 \cdot 2^{i-2}+1)^2 \equiv 9 \cdot 2^{2i-4}+3 \cdot 2^{i-1}+1$$
$$\equiv 3 \cdot 2^{(i+1)-2}+1 \pmod{2^{i+1}},$$

verifying the assertion.

For the second case,

$$5^{2^{(i+1)-4}} \equiv (2^i+3 \cdot 2^{i-2}+1)^2 \equiv 2^{2i}+3 \cdot 2^{2i-1}+9 \cdot 2^{2i-4}+2^{i+1}+3 \cdot 2^{i-1}+1$$
$$\equiv 3 \cdot 2^{(i+1)-2}+1 \pmod{2^{i+1}},$$

also verifying the assertion. Therefore by induction, $5^{2^{i-4}} \equiv 3 \cdot 2^{i-2}+1 \pmod{2^i}$ for $i \geq 5$.

From this congruence and because $5^n \equiv 1 \pmod 4$ we obtain

$$\sin\frac{2\pi 5^{n+2^{i-4}}}{2^i} = \sin\frac{2\pi 5^n(3\cdot 2^{i-2}+1)}{2^i}$$

$$= \sin\left[\frac{3\cdot 5^n\pi}{2} + \frac{2\pi 5^n}{2^i}\right]$$

$$= \sin\left[\frac{3\pi}{2} + \frac{2\pi 5^n}{2^i}\right]$$

$$= \sin\frac{3\pi}{2}\cos\frac{2\pi 5^n}{2^i} + \cos\frac{3\pi}{2}\sin\frac{2\pi 5^n}{2^i}$$

$$= -\cos\frac{2\pi 5^n}{2^i}.$$

A similar argument can be used to show that $\sin\dfrac{2\pi 5^n}{2^i} = \cos\dfrac{2\pi 5^{n+2^{i-4}}}{2^i}$.

Using these identities, (5.27) can be expressed as

$$-4j\sum_{n=0}^{2^{i-3}-1}\sin\frac{2\pi 5^n}{2^i}u^n = -4j\left[\sum_{n=0}^{2^{i-4}-1}\cos\frac{2\pi 5^{n+2^{i-4}}}{2^i}u^n - \sum_{n=0}^{2^{i-4}-1}\cos\frac{2\pi 5^n}{2^i}u^{n+2^{i-4}}\right]$$

$$\equiv 4ju^{2^{i-4}}\left[\sum_{n=0}^{2^{i-3}-1}\cos\frac{2\pi 5^n}{2^i}u^n\right] \pmod{u^{2^{i-3}}+1}. \tag{5.28}$$

Therefore the polynomials in (5.27) are just shifted versions of the polynomials in (5.26), and the two polynomial products could be carried out using the same algorithm and the same set of constants. The computation of the odd-indexed outputs of DFT(2^r) is thus equivalent to a union of polynomial products.

As was true for DFT(p^r), the residue reductions yielding zero in (5.24) are identical to the column sums that converted the computation of the even-indexed outputs into DFT(2^{r-1}). A nonsingular $2^r \times 2^r$ matrix exists that separates DFT(2^r) into a union of the polynomial products for the odd-indexed outputs and DFT(2^{r-1}) for the even-indexed outputs.

Theorem 5.3. *For an input vector of 2^r indeterminates, r a positive integer,* $\mu_B(\text{DFT}(2^r); Q) = 2^{r+1} - r - r^2 - 2.$

Proof. The system DFT(2^r) can be recursively decomposed into the union of a half-length DFT and a set of polynomial products modulo irreducible polynomials. The mth stage of this decomposition results in a pair of polynomial products modulo each of the polynomials $C_{2^i}(u) = u^{2^{i-1}} + 1$, $i = 1, 2, \ldots, m-2$. One polynomial multiplicand is always the same for every product modulo each of these irreducible polynomials.

Therefore DFT(2^r) is equivalent to

$$\bigcup_{i=3}^{r} \left[\bigcup_{j=1}^{i-2} \left[\bigcup_{m=0}^{1} C(C_{2^j}(u); f_j, l_{i,j,m}) \right] \right]$$

where the entries of f_j are the coefficients of $4 \sum\limits_{n=0}^{2^{j-1}-1} \cos\dfrac{2\pi 5^n}{2^j} u^n$, and the entries of $l_{i,j,m}$ are the coefficients of

$$\sum_{n=0}^{2^{i-2}-1} \sum_{m=0}^{1} x_{(2^{r-i}5^{-n}(-1)^m)_{2^r}} u^n v^m \pmod{u^{2^{i-3}}+1, v+(-1)^m}.$$

This system can be rewritten as

$$\bigcup_{j=1}^{r-2} \left[2(r-1-j) A_j(f_j) \right] M^{(j)} y$$

where $A_j(f_j)$ is the regular representation of f_j and $M^{(j)}$ contains the linear combination of the inputs specified by $l_{i,j,m}$.

This system is qrc reduced and $\dim L_Q(\bigcup\limits_{l \in L} r(f_l)) = \sum\limits_{l \in L} s_l \geq \sum\limits_{l \in L} (s_l - 1)$ for any subset L of $\{1, 2, \ldots, r-2\}$. The cardinality of y, excluding the modulo $u-1$ products, is equal to the sum of the degrees of the polynomial moduli, or

$$m = \sum_{i=1}^{r-2} 2(2^i - 1)$$

$$= 2^r - 2r.$$

Theorem 3.9 may now be applied to yield

$$\mu_B(\text{DFT}(2^r; Q)) \geq \sum_{j=1}^{r-2} 2(r-1-j)(2^j-1) + m$$

$$= 2m - \sum_{j=1}^{r-2} 2(r-1-j)$$

$$= 2(2^r - 2r) - 2(r-1)(r-2) + (r-1)(r-2)$$

$$= 2^{r+1} - 4r - r^2 + 3r - 2$$

$$= 2^{r+1} - r^2 - r - 2.$$

The algorithm of Theorem 3.3 achieves this lower bound, and therefore

$$\mu_B(\text{DFT}(2^r; Q)) = 2^{r+1} - r^2 - r - 2. \quad \blacksquare$$

Again, as for DFT(p^r), each of the subsystems computing the set of outputs with indexes whose greatest common divisor with 2^r is 2^i is equivalent to two length 2^{i-2} cyclic convolutions, excluding the factor $u-1$. Therefore DFT(2^r) is equivalent to the

system $\displaystyle\bigcup_{i=3}^{r} 2MPM\left[\dfrac{u^{2^{i-2}}-1}{u-1}\right]$.

5.5. Arbitrary Lengths

In this section the results of the previous three sections will be generalized to lengths that may include more than one distinct prime factor. Before proceeding with this analysis, it may be helpful to consolidate the results of the previous sections by expressing the multiplicative complexity of the DFT in a single formula for these lengths.

Using the decomposition of the DFT into cyclic convolutions, one for each divisor d of N, each cyclic convolution has length $\dfrac{\phi(d)}{\phi((d,4))}$. Each of the component polynomial multiplications requires $2 \cdot \deg P_i - 1$ m/d steps except for the multiplication modulo $u-1$ which is a rational multiplication requiring no m/d steps. There are two cyclic convolutions of the indicated length when d is a multiple of four, and one otherwise. The function $\phi((d,4))$ is equal to two for multiples of four and equal to one otherwise and will be used in accounting for this difference.

The result of Theorem 3.3 can now be applied. The number of irreducible polynomial factors of $u^{\phi(d)/\phi((d,4))} - 1$ is equal to the number of positive divisors of $\dfrac{\phi(d)}{\phi((d,4))}$, denoted by $\tau\left[\dfrac{\phi(d)}{\phi((d,4))}\right]$. Therefore

$$\mu_B(\text{DFT}(p^r); Q) = \sum_{d|p^r} \left[\phi((d,4))\left[2\left[\frac{\phi(d)}{\phi((d,4))}\right] - 1\right] - \tau\left[\frac{\phi(d)}{\phi((d,4))}\right] + 1\right]\right]$$

$$= \sum_{d|p^r} \left[2\phi(d) - \phi((d,4))\left[\tau\left[\frac{\phi(d)}{\phi((d,4))}\right]\right] + 1\right] \tag{5.29}$$

$$= 2p^r - \sum_{d|p^r} \phi((d,4))\left[\tau\left[\frac{\phi(d)}{\phi((d,4))}\right] + 1\right]$$

for any prime p (including 2) and positive integer r.

Unfortunately the expression in (5.29) is only valid for DFT lengths with a single distinct prime factor. To extend this result to other lengths it is first necessary to determine the difference in the structure of the DFT for composite lengths.

In the previous sections a necessary step in evaluating the multiplicative complexity was to come up with a method of generating the entire set of residues modulo N in a way that enabled the resulting computations to be recognized as a familiar system, a cyclic convolution. When N is composed of two or more distinct prime factors, an initial step toward obtaining a familiar system is the recognition that DFT(N) is the direct product $\mathrm{DFT}(N_1) \otimes \mathrm{DFT}(N_2) \otimes \cdots \otimes \mathrm{DFT}(N_m)$, where $N = \prod_{i=1}^{m} N_i$ and $(N_i, N_j) = 1$ for $i \neq j$.

The identification of the DFT as a direct product when the length is composed of relatively prime factors is clear from the prime factor index map, attributed to Thomas [38] and Good [20]. The prime factor index map [12] is simply an application of the Chinese remainder theorem for integers to the indexing of the inputs and outputs of a DFT. The Chinese remainder theorem for integers will now be stated to distinguish it from the version for polynomials presented in (3.2).

Let N_i, $i = 1, 2, \ldots, m$, be a set of integers such that $\prod_{i=1}^{m} N_i = N$, $N_i > 1$ for $i = 1, 2, \ldots, m$, and $(N_i, N_j) = 1$ for $i \neq j$. The Chinese remainder theorem for integers [26] states that any integer k $(0 \leq k < N)$ can be uniquely reconstructed from its residues modulo the m factors of N as

$$k \equiv \sum_{i=1}^{m} \langle k \rangle_{N_i} \frac{N}{N_i} n_i \ (\mathrm{mod}\, N),$$

where n_i satisfies the congruence $n_i \dfrac{N}{N_i} \equiv 1 \ (\mathrm{mod}\, N_i)$, $i = 1, 2, \ldots, m$.

The prime factor algorithm for length $N = \prod_{i=1}^{m} N_i$ is expressed as

$$
X_{\langle \sum_{i=1}^{m} k_i q_i \rangle_N} = \sum_{n_1=0}^{N_1-1} \sum_{n_2=0}^{N_2-1} \cdots \sum_{n_m=0}^{N_m-1} x_{\langle \sum_{i=1}^{m} n_i q_i \rangle_N} w_N^{(\sum_{i=1}^{m} n_i q_i)(\sum_{i=1}^{m} k_i q_i)}
$$

$$
= \sum_{n_1=0}^{N_1-1} \left[\sum_{n_2=0}^{N_2-1} \left[\cdots \left[\sum_{n_m=0}^{N_m-1} x_{\langle \sum_{i=1}^{m} n_i q_i \rangle_N} w_{N_m}^{n_m k_m} \right] \cdots \right] w_{N_2}^{n_2 k_2} \right] w_{N_1}^{n_1 k_1},
$$

(5.30)

for $k_i = 0, 1, \ldots, N_i-1$, and where q_i satisfies the congruences

$$q_i \equiv 1 \ (\mathrm{mod}\, N_i)$$

$$q_i \equiv 0 \ \left(\mathrm{mod}\, \frac{N}{N_i}\right)$$

for $i = 1, 2, \ldots, m$.

The index map used in (5.30) is not the only possible index map that can be used, but it does show that the DFT matrix decomposes into a direct product of the m blocks corresponding to the m sums in (5.30). The previous three sections have shown that these component DFT blocks may in turn be decomposed into a union of polynomial products modulo irreducible polynomials when the DFT length corresponding to the block is a power of a prime. Therefore, if the length N is expressed in its canonical factorization as a product of distinct primes, then the DFT of length N can be expressed as a direct product of polynomial multiplications modulo irreducible polynomials. This direct product is exactly the type of system that was analyzed in Theorem 3.10 as multivariate polynomial multiplication, with the exception that one polynomial does not have indeterminate coefficients. In evaluating the multiplicative complexity of this system, it is necessary to use the equivalence between direct products and direct sums of systems of polynomial products shown in Theorem 3.10 and then to analyze the resulting system using Theorem 3.9.

The following theorem uses this approach in deriving a formula for the multiplicative complexity of the one-dimensional DFT. Theorems 5.1, 5.2, and 5.3 can all be derived as special cases of this result.

Theorem 5.4. *If* $N = \prod_{i=1}^{m} p_i^{e_i}$, *where* p_i, $i = 1, 2, \ldots, m$, *are prime numbers such that* $p_i \neq p_j$ *for* $i \neq j$, *and* e_i, $i = 1, 2, \ldots, m$, *are positive integers, then*

$$\mu_B(\text{DFT}(N); Q) = 2N - \sum_{i_1=0}^{e_1} \sum_{i_2=0}^{e_2} \cdots \sum_{i_m=0}^{e_m} \phi\left(\left(\prod_{j=1}^{m} p_j^{i_j}, 4\right)\right)$$

$$\times \left[1 + \sum_{d_1 \mid \frac{\phi(p_1^{i_1})}{\phi((p_1^{i_1}, 4))}} \sum_{d_2 \mid \frac{\phi(p_2^{i_2})}{\phi((p_2^{i_2}, 4))}} \cdots \sum_{d_m \mid \frac{\phi(p_m^{i_m})}{\phi((p_m^{i_m}, 4))}} \frac{\prod_{k=1}^{m} \phi(d_k)}{\phi([d_1, d_2, \ldots, d_m])} \right].$$

Proof. It has been shown in the proof of Theorem 5.2 that $\text{DFT}(p^r)$ is equivalent to a set of cyclic convolutions, one for each divisor of the overall length, with the individual cyclic convolution lengths equal to the Euler function of the corresponding divisor. Similarly, Theorem 5.3 showed that $\text{DFT}(2^r)$ is equivalent to a set of cyclic convolutions, two for each divisor of the overall length greater than two, with lengths equal to one-fourth of the divisor. The divisors 1 and 2 contribute a rational multiplication each, which can be thought of as cyclic convolutions of length one.

For both $\text{DFT}(p^r)$ and $\text{DFT}(2^r)$ each cyclic convolution has a single rational multiplication, owing to the $u-1$ reduction. When the direct product $\overset{m}{\underset{i=1}{\otimes}} \text{DFT}(p_i^{e_i})$ is formed, the resulting system is a union of multidimensional cyclic convolutions. Each divisor of $N = \prod_{i=1}^{m} p_i^{e_i}$ will still yield one rational multiplication, corresponding to the multiplication modulo $u_1-1, u_2-1, \ldots, u_m-1$. From Lemma 5.1, the row rank of each multidimensional cyclic convolution is $\phi(d)-1$, where d is the divisor of N and $\phi(d)$ is thus the product of the cyclic convolution lengths in each dimension. Therefore the conditions of Theorem 3.9 are satisfied and the result of Corollary 3.2 can be applied to this system.

Converting the DFT into a set of multidimensional cyclic convolutions and applying Corollary 3.2 yields

$$
\mu_B(\text{DFT}(N); Q) = \sum_{i_1=0}^{e_1} \sum_{i_2=0}^{e_2} \cdots \sum_{i_m=0}^{e_m} \phi((\prod_{j=1}^{m} p_j^{i_j}, 4)) \left[\mu_B(\bigotimes_{j=1}^{m} MPM(u^{\phi(p_j^{i_j})/\phi((p_j^{i_j}, 4))} -1)) -1 \right]
$$

$$
= \sum_{i_1=0}^{e_1} \sum_{i_2=0}^{e_2} \cdots \sum_{i_m=0}^{e_m} \phi((\prod_{j=1}^{m} p_j^{i_j}, 4)) \left[2\prod_{j=1}^{m} \frac{\phi(p_j^{i_j})}{\phi((p_j^{i_j}, 4))} - 1 - \right.
$$

$$
\left. - \sum_{d_1 | \frac{\phi(p_1^{i_1})}{\phi((p_1^{i_1}, 4))}} \sum_{d_2 | \frac{\phi(p_2^{i_2})}{\phi((p_2^{i_2}, 4))}} \cdots \sum_{d_m | \frac{\phi(p_m^{i_m})}{\phi((p_m^{i_m}, 4))}} \frac{\prod_{k=1}^{m} \phi(d_k)}{\phi([d_1, d_2, \ldots, d_m])} \right]
$$

$$
= 2N - \sum_{i_1=0}^{e_1} \sum_{i_2=0}^{e_2} \cdots \sum_{i_m=0}^{e_m} \phi((\prod_{j=1}^{m} p_j^{i_j}, 4)) \left[1 + \right.
$$

$$
\left. + \sum_{d_1 | \frac{\phi(p_1^{i_1})}{\phi((p_1^{i_1}, 4))}} \sum_{d_2 | \frac{\phi(p_2^{i_2})}{\phi((p_2^{i_2}, 4))}} \cdots \sum_{d_m | \frac{\phi(p_m^{i_m})}{\phi((p_m^{i_m}, 4))}} \frac{\prod_{k=1}^{m} \phi(d_k)}{\phi([d_1, d_2, \ldots, d_m])} \right] . \blacksquare
$$

Theorem 5.4 gives the multiplicative complexity of the DFT for all possible one-dimensional sequence lengths. A set of Fortran routines is included in Appendix C that evaluate the formula of Theorem 5.4 for any sequence length up to $1531^2 = 2{,}343{,}961$ (if the largest integer constant permissible on the machine is greater than twice this number). Other limitations on this program are discussed in Appendix C.

Appendix D contains a tabulation of the complexity of the one-dimensional DFT for all lengths up to and including 1000. The program of Appendix C was used to compute the multiplicative complexity of the DFT for all lengths up to 50000. Table 5.1 summarizes this effort by showing the complexity for only those lengths for which the ratio of the required number of m/d steps to the sequence length is less than the same ratio computed for any longer sequence lengths. This choice of lengths is a special set of local minima for the function $\mu_B(\text{DFT}(N); Q)$ and these lengths are also likely to be very efficient for algorithms such as the Prime Factor Algorithm (PFA) [11, 20, 23] or Winograd Fourier Transform Algorithm (WFTA) [35, 42, 44] that are not minimal algorithms for longer lengths. The lengths greater than 50,000 may not be complete since the complexity was computed only for certain lengths that were considered to be likely candidates for inclusion.

Several aspects of the sequence lengths in Table 5.1 are interesting. The first notable feature is that the largest prime factor of the sequence length is almost monotonic with length (except the pairs of consecutive lengths, (60,72) and (504,720)). The primes 11 and 17 are not factors of any of the lengths in the table, but 13 and 19 are.

Both of these observations can be explained. Theorem 5.4 suggests that the complexity will generally be less when N has many divisors, implying that the best lengths will have only small prime factors. Theorem 5.4 also implies that better lengths have many prime factors shared between integers one less than each of the prime factors of N or between these integers and the other prime factors of N of multiplicity two or greater. An exception to this rule is a single shared factor of two.

Taking 11 as an example, 23 is the first prime that, when reduced by one, is divisible by 11. Since $11-1 = 10 = 2 \cdot 5$, a factor of 11 combines well with multiples of 25, which are also nonexistent in the table. Thus 11 is not likely to be a very efficient factor unless 23 and/or 25 are also factors of the sequence length. On the other hand, if unity is subtracted from 5, 7, 13, or 19, the resulting numbers have only 2 and 3 as prime factors, and thus share divisors with each other and with powers of 2 and 3. This is the reason that these primes are abundant as factors in the table.

The two exceptions to the monotonicity mentioned earlier both involve the prime 5 and likely are because $5-1 = 4$, and the extra savings mentioned only happen when 5 is multiplied by a power of 2 greater than $2^3 = 8$ or a different prime of the form $4k+1$, the smallest of which is 13. Subtracting unity from 17 yields 16 which, although seemingly quite composite, has fewer divisors than 12 or 18 and hence 17 is

N	$\mu(\text{DFT}(N);Q)$	μ/N	Factorization of N
4	0	0	2^2
8	2	.25	2^3
12	4	.333333	$2^2 \cdot 3$
24	12	.5	$2^3 \cdot 3$
48	38	.791667	$2^4 \cdot 3$
60	56	.933333	$2^2 \cdot 3 \cdot 5$
72	70	.972222	$2^3 \cdot 3^2$
120	120	1	$2^3 \cdot 3 \cdot 5$
240	274	1.141667	$2^4 \cdot 3 \cdot 5$
360	460	1.277778	$2^3 \cdot 3^2 \cdot 5$
504	660	1.309524	$2^3 \cdot 3^2 \cdot 7$
720	986	1.369444	$2^4 \cdot 3^2 \cdot 5$
840	1152	1.371429	$2^3 \cdot 3 \cdot 5 \cdot 7$
1008	1422	1.410714	$2^4 \cdot 3^2 \cdot 7$
1680	2426	1.444048	$2^4 \cdot 3 \cdot 5 \cdot 7$
2520	3720	1.476190	$2^3 \cdot 3^2 \cdot 5 \cdot 7$
5040	7722	1.532143	$2^4 \cdot 3^2 \cdot 5 \cdot 7$
6552	10272	1.567766	$2^3 \cdot 3^2 \cdot 7 \cdot 13$
10920	17520	1.604396	$2^3 \cdot 3 \cdot 5 \cdot 7 \cdot 13$
13104	21078	1.608516	$2^4 \cdot 3^2 \cdot 7 \cdot 13$
21840	35626	1.631227	$2^4 \cdot 3 \cdot 5 \cdot 7 \cdot 13$
32760	53520	1.633700	$2^3 \cdot 3^2 \cdot 5 \cdot 7 \cdot 13$
65520	108594	1.657418	$2^4 \cdot 3^2 \cdot 5 \cdot 7 \cdot 13$
131040	224718	1.714881	$2^5 \cdot 3^2 \cdot 5 \cdot 7 \cdot 13$
196560	347238	1.766575	$2^4 \cdot 3^3 \cdot 5 \cdot 7 \cdot 13$
207480	373200	1.798728	$2^3 \cdot 3 \cdot 5 \cdot 7 \cdot 13 \cdot 19$
311220	561120	1.802969	$2^2 \cdot 3^2 \cdot 5 \cdot 7 \cdot 13 \cdot 19$
622440	1122336	1.803123	$2^3 \cdot 3^2 \cdot 5 \cdot 7 \cdot 13 \cdot 19$

Table 5.1. Number of m/d steps necessary to compute DFT(N) over Q for real inputs for selected lengths.

not found as a factor in Table 5.1. A restricted sampling suggests that the shortest length satisfying the conditions of Table 5.1 and divisible by 17 is 10,581,480. No length was found with a divisor of 11 satisfying the conditions of Table 5.1 using a limited search up to $N = 10^8$. Perhaps such a length exists for a much larger N.

Several examples will now be worked out to show how Theorem 5.4 can be applied to obtain less complicated formulas for the multiplicative complexity of DFTs of certain composite lengths. Many unit step functions occur in these formulas because of the steplike nature of the function in the innermost sum. The unit step function (or sequence) is defined by

$$u(n) = \begin{cases} 0, & n < 0 \\ 1, & n \geq 0 \end{cases}$$

and in the following examples the identity $u(n) = 1 - \sum_{i=0}^{n-1} \delta(i)$, where

$$\delta(n) = \begin{cases} 0, & n \neq 0 \\ 1, & n = 0 \end{cases}$$

is the unit sample sequence or discrete time impulse [28], will frequently be used to simplify the resulting formulas.

Example 5.1. Consider the system $DFT(2^a p^b)$ where p is an odd prime and a and b are integers. The multiplicative complexity of this system is determined in the following derivation using Theorem 5.4.

The prime p may be expressed as $p = 2^k(2l+1)+1$, where k is the largest power of two dividing $p-1$ and $2l+1$ is the largest odd integer dividing $p-1$. The unity term inside the sums in Theorem 5.4 can be moved outside the sums, yielding a term equal to $\tau(N) + u(a-2)\tau(N/(2^a, 4))$, where $\tau(N)$ is the number of divisors of N and $u(n)$ is the unit step sequence. From Theorem 5.4 we obtain

$$\mu_B(DFT(2^a p^b); Q) = 2^{a+1} p^b - (a+1)(b+1) - (a-1)(b+1)u(a-2)$$

$$- \sum_{i=0}^{a} \sum_{j=0}^{b} \sum_{d_1 \mid \frac{\phi(2^i)}{\phi(2^i,4)}} \sum_{d_2 \mid \phi(p^j)} \phi((2^i, 4))\phi((d_1, d_2))$$

$$= 2^{a+1} p^b - (a+1)(b+1) - (1-\delta(a))(a-1)(b+1)$$

$$- (2-\delta(a)) \left[1 + \sum_{j=1}^{b} \sum_{d_2 \mid 2^k(2l+1)p^{j-1}} 1 \right] - (1-\delta(a)-\delta(a-1))$$

$$\times \left[\sum_{i=2}^{a} \sum_{d_1 \mid 2^{i-2}} 1 + \sum_{i=2}^{a} \sum_{j=1}^{b} \sum_{d_1 \mid 2^{i-2}} \sum_{d_2 \mid 2^k(2l+1)p^{j-1}} 2\phi(d_1, d_2) \right]$$

$$= 2^{a+1}p^b - 2a(b+1) - 2 - 2\sum_{j=1}^{b}(k+1)\tau(2l+1)j$$

$$-\sum_{i=2}^{a}(i-1) - \sum_{i=2}^{a}\sum_{j=1}^{b}(k+1)\tau(2l+1)j$$

$$-\sum_{i=2}^{a}\sum_{j=1}^{b}\sum_{m=0}^{i-2}\sum_{n=0}^{k}2\tau(2l+1)j\phi((2^m, 2^n))$$

$$+\delta(a)\left[(a-1)(b+1)+1+\sum_{j=1}^{b}(k+1)\tau(2l+1)j\right]$$

$$= 2^{a+1}p^b - 2ab - a^2 - a - 2 - (b^2+b)(k+1)\tau(2l+1)$$

$$-(b^2+b)\tau(2l+1)\sum_{i=2}^{a}\left[1 + \sum_{m=1}^{i-2}\sum_{n=1}^{k}\min(2^{m-1}, 2^{n-1})\right.$$

$$\left.+\sum_{n=1}^{k}1 + \sum_{m=1}^{i-2}1\right] + \delta(a)\left[-b+\tfrac{1}{2}(b^2+b)(k+1)\tau(2l+1)\right]$$

$$= 2^{a+1}p^b - 2ab - a^2 - a - 2 - (b^2+b)(k+1)\tau(2l+1)$$

$$-(b^2+b)\tau(2l+1)\sum_{i=2}^{a}\left[i+k-1+\sum_{m=1}^{\min(i-2,k)}\left(\sum_{n=1}^{m}2^{n-1}\right.\right.$$

$$\left.\left.+\sum_{n=m+1}^{\max(i-2,k)}2^{m-1}\right)\right] + \delta(a)\left[-b+\tfrac{1}{2}(b^2+b)(k+1)\tau(2l+1)\right]$$

$$= 2^{a+1}p^b - 2ab - a^2 - a - 2 - (b^2+b)(k+1)\tau(2l+1)$$

$$-(b^2+b)\tau(2l+1)\sum_{i=2}^{a}\left[i+k-1+\sum_{m=1}^{\min(i-2,k)}(2^m-1\right.$$

$$\left.+(\max(i-2,k)-m)2^{m-1})\right] + \delta(a)\left[-b+\tfrac{1}{2}(b^2+b)(k+1)\tau(2l+1)\right]$$

$$= 2^{a+1}p^b - 2ab - a^2 - a - 2 - (b^2+b)(k+1)\tau(2l+1)$$

$$- (b^2+b)\tau(2l+1)\sum_{i=2}^{a}\left[(|i-2-k|+3)2^{\min(i-2,\,k)} - 2\right]$$

$$+ \delta(a)\left[-b + \tfrac{1}{2}(b^2+b)(k+1)\tau(2l+1)\right]$$

$$= 2^{a+1}p^b - 2ab - a^2 - a - 2 - (b^2+b)\tau(2l+1)(k-2a+3)$$

$$- (b^2+b)\tau(2l+1)\left[\sum_{i=2}^{\min(k+2,\,a)}(k-i+5)2^{i-2} + \sum_{i=k+3}^{a}(i-k+1)2^{k}\right]$$

$$+ \delta(a)\left[-b - \tfrac{1}{2}(k+5)(b^2+b)\tau(2l+1)\right]$$

$$- \delta(a-1)(k+5)(b^2+b)\tau(2l+1)$$

$$= 2^{a+1}p^b - 2ab - a^2 - a - 2$$

$$- (b^2+b)\tau(2l+1)(2^{k-1}(a^2+k^2-2ak-3a-3k+6)-2a-2)$$

$$+ \delta(a)\left[-b + (2^{k-1}(k^2-3k+6) - \frac{k+5}{2})(b^2+b)\tau(2l+1)\right]$$

$$+ \delta(a-1)\left[2^{k-1}(k^2-5k+10)-k-5\right](b^2+b)\tau(2l+1)$$

$$+ \sum_{i=2}^{k-1}\delta(a-i)\left[2^{k-1}(k^2+i^2-2ik+3i-3k+6)-2^{i-1}(6-i+k)\right]$$

$$\times (b^2+b)\tau(2l+1).$$

Note that the upper limit on the sum of the delta functions is $k-1$ rather than $k+2$ because the terms of the sum are identically zero for $k \le i \le k+3$. The resulting expression for the multiplicative complexity of DFT($2^a p^b$) seems complicated, but is simple to evaluate to obtain complexities of this type. We easily obtain

$$\mu_B(\text{DFT}(2^a 3^b); Q) = 2^{a+1} 3^b - (a^2 + a + 2)(b^2 + b + 1) + b^2(2a + \delta(a))$$

and

$$\mu_B(\text{DFT}(2^a 5^c); Q) = 2^{a+1} 5^c - (a^2 - 2a + 3)(2c^2 + 2c + 1) - 2ac - 3a + 1$$
$$+ \tfrac{1}{2}(9c^2 + 7c)\delta(a) + (c^2 + c)\delta(a-1).$$

5.6. DFTs with Complex-Valued Inputs

The preceding analysis of the multiplicative complexity of the DFT has provided formulas for the number of m/d steps necessary and sufficient to compute DFTs of arbitrary length with real-valued inputs. The extension of this analysis to DFTs with complex-valued inputs can be done in several ways. One interpretation of this extension will be presented to show that for all lengths exactly twice the number of m/d steps are required for a DFT with complex-valued input as for a DFT with real-valued input.

In the analysis of the DFT of lengths that are powers of primes it was shown that the cyclic convolutions imbedded in the DFT are all of even length and thus can be decomposed into products modulo $u^{r/2} - 1$ and $u^{r/2} + 1$ where r is the cyclic convolution length. This decomposition resulted in a union of systems of purely real and purely imaginary polynomial products. This same decomposition also produces a union of purely real and purely imaginary polynomial products for DFTs of general length since they can be represented as direct products of the same cyclic convolutions encountered for prime-power lengths.

Except for lengths that are multiples of four, the real and imaginary submatrices of this decomposition of the DFT are independent. Therefore, including an imaginary component in the input forms the union of two identical systems, each equivalent to the original DFT with real-valued input. The resulting complexity, according to Theorem 3.9, is exactly double that of the DFT with real-valued input.

When the length is a multiple of four, then the real and imaginary submatrices are identical, but they occur in a union that is qrc reduced and thus the multiplicative complexity is again exactly double that of the DFT with real-valued input.

5.7. Multidimensional DFTs

The original intent of this book was to present the multiplicative complexity of only the one-dimensional DFT. However, in the interest of completeness, one known result will be presented for multidimensional DFTs. From the discussion of the prime factor algorithm, it is already clear that a one-dimensional DFT of composite length is equivalent to a multidimensional DFT having the same number of samples, with the lengths in each dimension pairwise relatively prime. We will now investigate a multidimensional DFT in which factors are shared between lengths in the different dimensions. This type of system is not equivalent to a single one-dimensional DFT.

Auslander *et al.* [3] have demonstrated that the two-dimensional DFT of length p in each dimension, p a prime, is equivalent to the union of $p+1$ DFT(p) cores and a product by unity. This decomposition is derived by interpreting the pairs of indexes as elements of $GF(p^2)$ which may be generated by raising a primitive root to successive integer powers, in direct analogy to Rader's permutation of the one-dimensional prime-length DFT. This idea may be generalized to multiple dimensions, by considering the k-tuples of indexes as elements of $GF(p^k)$.

In two dimensions the multiplicative complexity of this system is

$$\mu_B(\text{DFT}(p:2); Q) = (p+1)(2p-\tau(p-1)-3).$$

One way of generalizing this to k dimensions is by realizing that the two-dimensional system is the direct product (by itself) of the direct sum of a length-1 and length-$(p-1)$ cyclic convolution. This yields a length-1 cyclic convolution, two length-$(p-1)$ cyclic convolutions, and a two-dimensional $(p-1)$ by $(p-1)$ cyclic convolution. Obviously the $(p-1)$ by $(p-1)$ two-dimensional cyclic convolution is equivalent to $p-1$ length-$(p-1)$ one-dimensional cyclic convolutions. The k-dimensional DFT is simply the direct product of the $(k-1)$-dimensional DFT with the one-dimensional DFT. Let m_k be the number of one-dimensional length-$(p-1)$ cyclic convolutions in the decomposition of DFT$(p:k)$. Each of the m_{k-1} blocks of the $(k-1)$-dimensional DFT combine with the similar one-dimensional block to form $p-1$ similar blocks in the k-dimensional DFT. In addition to these $(p-1)m_{k-1}$ blocks, there will be m_{k-1} similar blocks formed by the product with the unity block in the one-dimensional DFT and one block formed by the product of the length-$(p-1)$ cyclic convolution block in the one-dimensional DFT with the unity block in the $(k-1)$-dimensional DFT. A recursion formula for the number of blocks is therefore

$$m_k = (p-1)m_{k-1}+m_{k-1}+1$$
$$= pm_{k-1}+1$$

whose solution is

$$m_k = \frac{p^k-1}{p-1}.$$

Therefore,

$$\mu_B(\text{DFT}(p:k); Q) = \frac{p^k-1}{p-1}(2p-\tau(p-1)-3).$$

This complexity is slightly less than that obtained by evaluating the complexity of the system as that of a direct product of cyclic convolutions, even taking into account the rational multiplications. No similar result has yet been reported for the case when the prime p is raised to a power greater than one in one or more of the dimensions.

5.8. Summary of Chapter 5

After introducing the discrete Fourier transform, this chapter concentrated on deriving the multiplicative complexity of the one-dimensional DFT for all possible sequence lengths. This analysis began with prime lengths and then was extended to power-of-prime lengths, power-of-two lengths, and finally to arbitrary lengths. A brief final section discussed the multiplicative complexity of the DFT for complex-valued inputs.

The first conclusion to be drawn is that the number of m/d steps required to compute a DFT of any length is always less than twice the sequence length when the input sequence is real-valued. The exact number of m/d steps required to compute a DFT of a given length is a complicated function depending on the factorization of the sequence length N, the factorization of the integers one less than the prime divisors of N, and how these factors are duplicated from one factorization to another.

It is possible to simplify the general formula for the multiplicative complexity for sequence lengths that have common features in their factorizations. An example is explicitly worked out for sequence lengths $N = 2^a p^b$, which may then be applied to any odd prime p. The multiplicative complexity of the DFT for complex-valued input is exactly twice that of the DFT with real-valued input for any sequence length.

The multiplicative complexity of a multidimensional DFT is analyzed for systems in which the length is identical, and prime, in each dimension. Extension of this result to (possibly differing) powers of an identical prime in the various dimensions would permit the multiplicative complexity of the multidimensional DFT to be computed for any arbitrary system.

CHAPTER 6

Restricted and Constrained DFTs

The theory of the multiplicative complexity of the DFT will now be developed from a different point of view that generalizes the results obtained for the complete DFT to cases where it is only desired to compute a portion of the DFT outputs, but not the entire set. Such a case might arise when certain frequencies or frequency bands are being monitored for changes in energy, whereas other frequencies or frequency bands are always ignored. This includes the case of FFT output pruning in which the spectrum is only needed in a narrow passband about DC.

The remainder of the chapter considers DFTs that have constraints on the inputs. Examples of this include FFT input pruning, one example of which is the zoom-FFT [49] that computes a long DFT from a small number of samples to give an impression of a smooth spectrum, primarily for plotting purposes.

6.1. Restricting DFT Outputs to One Point

This analysis begins with the determination of the number of multiplications necessary to compute a single DFT output for a given sequence length. Clearly, the multiplicative complexity is dependent both on the sequence length and on the index of the desired sample since, for example, the computation of the zero-frequency (or DC) term never requires any multiplications. Theorem 2.5 will be extensively used in the derivation of the multiplicative complexity of the computation of a single DFT output.

We first define $DFT(N;k)$ as the system that computes the kth output of a length-N DFT. Let (N,k) denote the greatest common divisor of N and k, then we can show that $DFT(N;k)$ is equivalent to the system $DFT(N/(N,k);1)$. Let $k_1 = k/(N,k)$ and $N_1 = N/(N,k)$; both k_1 and N_1 are positive integers. Now

$$X_k = \sum_{n=0}^{N-1} x_n e^{-j2\pi nk/N} = \sum_{n=0}^{N-1} x_n e^{-j2\pi nk_1/N_1}$$
$$= \sum_{n_1=0}^{N_1-1} (\sum_{n_2=0}^{(N,k)-1} x_{n_1+N_1 n_2} e^{-j2\pi n_1 k_1/N_1}) \tag{6.1}$$

showing that $DFT(N;k)$ is equivalent to $DFT(N_1;k_1)$ where $(N_1,k_1) = 1$. Since N_1 and k_1 are relatively prime, multiples of k_1 generate a complete residue system modulo N_1, therefore for each term in the sum over n_1 in (6.1) there is a term with the identical power of $e^{-j2\pi/N_1}$ in a similar expression for DFT(N_1; 1), hence the systems DFT(N_1; 1) and DFT(N_1; k_1) are identical except for a permutation of the inputs.

The problem of computing the kth output of DFT(N) can be thought of as the evaluation at the point $e^{-j2\pi k/N}$ of the $N-1^{st}$ degree polynomial whose coefficients are equal to the input values. This system differs from polynomial evaluations considered by early researchers in multiplicative complexity theory in that they knew the polynomial coefficients in advance and wanted to evaluate that fixed polynomial at any given point. For DFT(N; k), the problem is to evaluate an arbitrary polynomial of known degree at a fixed point. The application of Horner's rule to this computation is also known as a first-order Goertzel algorithm [18] and requires approximately $4N$ real multiplications to carry out. A second-order Goertzel algorithm requires $2N$ real multiplications, which is still more than twice as many multiplications as the algorithms that will now be derived.

We will first consider the case DFT(N; 1) since we have shown that all possible choices of N and k are equivalent to a system of this form. The column rank (over Q) of the row vector multiplying the column vector of inputs must be determined to apply Theorem 2.5 to this system. It will be assumed that the inputs are arbitrary real numbers, and the fact that the output DFT sample is really a complex number will be dealt with after dispensing with some algebraic relationships that are more easily expressed in complex exponential form.

For simplicity of notation let $w = e^{-j2\pi/N}$. The row vector of interest is $a = [w^0 \ w^1 \ w^2 \ \cdots \ w^{N-1}]$. The number of independent rational linear combinations of the powers of w that are rational themselves (i.e., are congruent to zero in the quotient field $Q(w)/Q$) must be determined to obtain the column rank of a over Q. This is identical to the problem of Corollary 5.2 except that the real and imaginary parts are not combined, yielding a column rank of $\phi(N)-\phi((N,4))$ for this row of the DFT matrix. We are now prepared to state the following theorem.

Theorem 6.1. $\mu_B(\text{DFT}(N; 1); Q) = \phi(N)-\phi((N,4))$ *for real-valued inputs.*

Proof. The proof follows from the preceding remarks and Theorem 2.5. ∎

Corollary 6.1. $\mu_B(\text{DFT}(N; k); Q) = \phi(\dfrac{N}{(N,k)})-\phi((\dfrac{N}{(N,k)},4))$ *for real-valued inputs.*

Proof. This corollary follows immediately from the theorem and (6.1). ∎

When the input is complex-valued, then each multiplication is replaced by a system of polynomial multiplication modulo u^2+1 and the result of Corollary 5.2 can be directly applied. When the length is not divisible by four, then the system decomposes into the union of the computation of the component due to the real part of the input sequence and the component due to the imaginary part. The identity

$$X_1 = \sum_{n=0}^{N-1} (x_n + jy_n)e^{-j2\pi n/N}$$

$$= \sum_{n=0}^{N-1} \left[x_n e^{-j2\pi n/N} + y_n e^{-j2\pi(n+1/4N)/N} \right]$$

$$= \sum_{n=0}^{1/4N-1} (x_n + y_{n+3/4N})e^{-j2\pi n/N} + \sum_{n=1/4N}^{N-1} (x_n + y_{n-1/4N})e^{-j2\pi n/N}$$

shows that for multiples of four, evaluation of DFT(N; 1) for complex inputs is identical to the evaluation for real inputs and thus requires the same number of m/d steps.

Theorem 6.2. $\mu_B(\text{DFT}(N; 1); Q) = 3\phi(N) - \phi((N, 4))\phi(N) - 2$ *for complex-valued inputs.*

Proof. The complexity will be $\phi(N) - 2$ when $4 \mid N$ and will be $2(\phi(N) - 1) = 2\phi(N) - 2$ when $4 \nmid N$. The formula stated in the theorem uses the function $3 - \phi((N, 4))$ to obtain these coefficients of $\phi(N)$. ∎

Corollary 6.2. $\mu_B(\text{DFT}(N; k); Q) = 3\phi(\dfrac{N}{(N, k)}) - \phi((\dfrac{N}{(N, k)}, 4))\phi(\dfrac{N}{(N, k)}) - 2$ *for complex-valued inputs.*

Proof. As before, this corollary follows immediately from the preceding theorem and (6.1). ∎

The number of multiplications required to compute DFT(N; 1) is summarized for all values of N up to 102 in Table 6.1.

6.2. Constraining DFT Inputs to One Point

Suppose that only one input to a DFT was non-zero. The analysis of the multiplicative complexity of this system is trivial since the Φ matrix of this system is exactly the transpose of that considered in the analysis of the restriction of the DFT output to a single point. Therefore, by Theorem 2.4, the complexity is exactly the same as that of DFT(N; k), for the input point n sharing the same greatest common divisor with N as k. The following corollary is thus a trivial result.

Corollary 6.3. $\mu_B(\text{DFT}(N; n); Q) = \phi(\dfrac{N}{(N, n)}) - \phi((\dfrac{N}{(N, n)}, 4))$ *for real-valued input* x_n.

6.3. DFTs with Symmetric Inputs

A special type of input that commonly needs to be transformed with a DFT is a sequence with either even or odd symmetry. These systems could also be identified as those which restrict the output of a DFT of real-valued inputs to the real part or imaginary part only. Oddly enough, these types of system have already been encountered in the process of determining the multiplicative complexity of the DFT, and arise naturally in this analysis for all lengths with only one distinct prime divisor. The system of (5.6) is exactly a system that computes the DFT of a symmetric sequence.

N	μ_r	μ_c	N	μ_r	μ_c	N	μ_r	μ_c
1	0	0	35	23	46	69	43	86
2	0	0	36	10	10	70	23	46
3	1	2	37	35	70	71	69	138
4	0	0	38	17	34	72	22	22
5	3	6	39	23	46	73	71	142
6	1	2	40	14	14	74	35	70
7	5	10	41	39	78	75	39	78
8	2	2	42	11	22	76	34	34
9	5	10	43	41	82	77	59	118
10	3	6	44	18	18	78	23	46
11	9	18	45	23	46	79	77	154
12	2	2	46	21	42	80	30	30
13	11	22	47	45	90	81	53	106
14	5	10	48	14	14	82	39	78
15	7	14	49	41	82	83	81	162
16	6	6	50	19	38	84	22	22
17	15	30	51	31	62	85	63	126
18	5	10	52	22	22	86	41	82
19	17	34	53	51	102	87	55	110
20	6	6	54	17	34	88	38	38
21	11	22	55	39	78	89	87	174
22	9	18	56	22	22	90	23	46
23	21	42	57	35	70	91	71	142
24	6	6	58	27	54	92	42	42
25	19	38	59	57	114	93	59	118
26	11	22	60	14	14	94	45	90
27	17	34	61	59	118	95	71	142
28	10	10	62	29	58	96	30	30
29	27	54	63	35	70	97	95	190
30	7	14	64	30	30	98	41	82
31	29	58	65	47	94	99	59	118
32	14	14	66	19	38	100	38	38
33	19	38	67	65	130	101	99	198
34	15	30	68	30	30	102	31	62

Table 6.1. Number of multiplications necessary to compute DFT(N; 1) for real and complex inputs.

Repeating (5.6), we have

$$\sum_{i=0}^{p-2} \overline{w}_i u^i \equiv 2 \sum_{i=0}^{(p-3)/2} (\cos \frac{2\pi g^i}{p} -1)u^i \quad (\text{mod } u^{(p-1)/2}-1)$$

for prime lengths. Thus a prime-length DFT with symmetric input, denoted by $\text{DFT}_s(p)$, is equivalent to a cyclic convolution of length $\frac{p-1}{2}$. The single rational multiplication found in the prime-length DFT is always in this symmetric part since the factor $u-1$ always divides $u^{(p-1)/2}-1$.

Corollary 6.4. $\mu_B(\text{DFT}_s(p); Q) = p - \tau \left[\frac{p-1}{2} \right] - 2.$

Proof. The assumptions necessary to prove this result have already been stated in Theorem 5.1. The equivalence of $\text{DFT}_s(p)$ to a cyclic convolution yields

$$\mu_B(\text{DFT}_s(p); Q) = \mu_B \left[MPM \left[\frac{u^{(p-1)/2}-1}{u-1} \right] \right]$$

$$= 2 \left[\frac{p-1}{2} \right] - \tau \left[\frac{p-1}{2} \right] - 1$$

$$= p - \tau \left[\frac{p-1}{2} \right] - 2. \quad \blacksquare$$

The prime-length DFT with odd symmetric (or antisymmetric) input, denoted by $\text{DFT}_a(p)$ is partially described by the polynomial in (5.7) and is equivalent to a polynomial product modulo $u^{(p-1)/2}+1$. Actually, since the system $\text{DFT}(p)$ decomposes into the union $\text{DFT}_s(p) \cup \text{DFT}_a(p)$, the multiplicative complexity of $\text{DFT}_a(p)$ is exactly the difference between the complexities of $\text{DFT}(p)$ and $\text{DFT}_s(p)$.

Corollary 6.5. $\mu_B(\text{DFT}_a(p); Q) = p - \tau(p-1) + \tau \left[\frac{p-1}{2} \right] - 1.$

Proof. The assumptions of Theorem 5.1 still hold, thus

$$\mu_B(\text{DFT}_a(p); Q) = \mu_B(\text{DFT}(p); Q) - \mu_B(\text{DFT}_s(p); Q)$$

$$= 2p - \tau(p-1) - 3 - (p - \tau \left[\frac{p-1}{2} \right] - 2)$$

$$= p - \tau(p-1) + \tau \left[\frac{p-1}{2} \right] - 1. \quad \blacksquare$$

The results of Corollaries 6.4 and 6.5 can be trivially extended to sequence lengths that are powers of odd primes and powers of two. The extension to arbitrary integers is not as straightforward as for the general DFT since the prime factor decomposition of the symmetric DFT yields both symmetric and antisymmetric components in the direct product. Before investigating this problem, though, lengths that are odd prime powers or powers of two will be investigated.

Corollary 6.6. $\mu_B(\mathrm{DFT}_s(p^r); Q) = p^r - r - 1 - \dfrac{r^2 + r}{2} \tau \left\lfloor \dfrac{p-1}{2} \right\rfloor.$

Proof. The assumptions of Theorem 5.2 hold, thus

$$\mu_B(\mathrm{DFT}_s(p^r); Q) = \mu_B \left[\bigcup_{i=1}^{r} MPM \left[\frac{u^{(p^i - p^{i-1})/2} - 1}{u - 1} \right]; Q \right]$$

$$= \sum_{i=1}^{r} \left[\frac{2(p^i - p^{i-1})}{2} - \tau \left[\frac{p^i - p^{i-1}}{2} \right] - 1 \right]$$

$$= p^r - r - 1 - \frac{r^2 + r}{2} \tau \left\lfloor \frac{p-1}{2} \right\rfloor. \quad \blacksquare$$

As for prime lengths, the computation of $\mathrm{DFT}(p^r)$ for an antisymmetric sequence is exactly the computation remaining when computing the full DFT after taking care of the symmetric part. The following corollary is obvious from this remark.

Corollary 6.7. $\mu_B(\mathrm{DFT}_a(p^r); Q) = p^r - 1 - \dfrac{r^2 + r}{2} \left[\tau(p-1) - \tau \left\lfloor \dfrac{p-1}{2} \right\rfloor \right].$

The results for prime-length sequences are the special case $r = 1$ of the results for sequence lengths that are powers of primes. For the sequence lengths considered up to this point, the multiplicative complexity for symmetric inputs is always less than that of antisymmetric inputs. This is easily demonstrated by determining the relationship between $\tau(p-1)$ and $\tau \left\lfloor \dfrac{p-1}{2} \right\rfloor$.

All odd primes can be expressed in the form $p = 2^k(2l+1) + 1$, where k and l are positive integers. The factors 2^k and $2l+1$ of $p-1$ are relatively prime and since $\tau(n)$ is a multiplicative function it follows that $\tau(p-1) = \tau(2^k)\tau(2l+1) = (k+1)\tau(2l+1)$. Using this same idea and the knowledge that $k \geq 1$ yields $\tau \left\lfloor \dfrac{p-1}{2} \right\rfloor = \tau(2^{k-1})\tau(2l+1) = k\tau(2l+1).$

Substitution of the expressions for $\tau(p-1)$ and $\tau \left\lfloor \dfrac{p-1}{2} \right\rfloor$ into the formulas of

Corollaries 6.6 and 6.7 and taking the difference yields

$$\mu_B(\text{DFT}_a(p^r);Q)-\mu_B(\text{DFT}_s(p^r);Q) = r+\frac{r^2+r}{2}(k-1)\tau(2l+1). \tag{6.2}$$

Since $r \geq 1$, $k \geq 1$, and the function $\tau(n)$ is always positive, the first term of the sum in (6.2) is always positive and the second is always nonnegative, hence $\mu_B(\text{DFT}_a(p^r);Q) > \mu_B(\text{DFT}_s(p^r);Q)$.

From (6.2) it is apparent that for primes of the form $p = 4k+3$, the difference between the multiplicative complexities of $\text{DFT}_s(p^r)$ and $\text{DFT}_a(p^r)$ is exactly the number of rational multiplications present in $\text{DFT}(p^r)$. This interesting development can be traced to the factorizations of $u^{(p-1)/2}-1$ and $u^{(p-1)/2}+1$ for these lengths, i.e.,

$$u^{(p-1)/2}-1 = \prod_{d|(2l+1)} C_d(u)$$

$$u^{(p-1)/2}+1 = \prod_{d|(2l+1)} C_{2d}(u) = \prod_{d|(2l+1)} C_d(-u)$$

where $p = 2 \cdot (2l+1)+1$. Therefore the polynomial multiplication modulo $u^{(p-1)/2}+1$ is equivalent to a cyclic convolution of length $(p-1)/2$. This equivalence causes the DFT computation for these lengths to be more regular and was originally exploited by Chu and Burrus [13].

When the sequence length is a power of two, then the even and odd symmetric parts of the input yield identical systems, as was demonstrated in (5.27) and (5.28). Since the system $\text{DFT}(2^r)$ decomposes into a union of $\text{DFT}_s(2^r)$ and $\text{DFT}_a(2^r)$ that are identical, the multiplicative complexity of each of the components must be exactly half of $\mu_B(\text{DFT}(2^r);Q)$. The following corollary follows immediately from these remarks and is stated without proof.

Corollary 6.8. $\mu_B(\text{DFT}_s(2^r);Q) = \mu_B(\text{DFT}_a(2^r);Q) = 2^r - \frac{r^2+r}{2} - 1.$

Extending these results to general lengths is more complicated than for the full DFT since the direct products formed in the prime factor decomposition include both symmetric and antisymmetric components. When the length is a multiple of four, the frequency of each symmetric and antisymmetric component is identical for either symmetric or antisymmetric inputs, and therefore the multiplicative complexity of each is exactly half that of the full DFT. A formula for the multiplicative complexity of the symmetric or antisymmetric DFT would be much more complicated than that of the full DFT. A method will instead be stated by which these complexities can be computed for any given sequence length.

As just mentioned, for N a multiple of four, $\mu_B(\text{DFT}_a(N);Q) = \mu_B(\text{DFT}_s(N);Q) = \frac{1}{2}\mu_B(\text{DFT}(N);Q)$. When $(N, 4) = 2$ then $\mu_B(\text{DFT}_s(N);Q) = 2\mu_B(\text{DFT}_s(N/2);Q)$ and $\mu_B(\text{DFT}_a(N);Q) = 2\mu_B(\text{DFT}_a(N/2);Q)$.

Thus all that remains is to determine these complexities for odd sequence lengths composed of more than one prime factor.

Since the computation of the full DFT breaks down into a direct product of cyclic convolutions which then break down into symmetric and antisymmetric components, we simply need to extract only those portions of the full DFT contributing to the real (or imaginary) part of the output. As an example, for $N = p_1^{e_1} p_2^{e_2}$ having two prime factors, we have the following equivalences:

$$\text{DFT}_s(N) = \text{DFT}_s(p_1^{e_1}) \otimes \text{DFT}_s(p_2^{e_2}) \cup \text{DFT}_a(p_1^{e_1}) \otimes \text{DFT}_a(p_2^{e_2})$$

and

$$\text{DFT}_a(N) = \text{DFT}_s(p_1^{e_1}) \otimes \text{DFT}_a(p_2^{e_2}) \cup \text{DFT}_a(p_1^{e_1}) \otimes \text{DFT}_s(p_2^{e_2}).$$

This system does not decompose into unions of multidimensional cyclic convolutions since the DFT_a portions are skew-cyclic convolutions (products modulo u^r+1 rather than u^r-1).

Clearly a DFT of any odd length can be decomposed in this way into a union of direct products of cyclic and skew-cyclic convolutions. For odd-length symmetric DFTs one simply decomposes the full DFT in this manner, selecting only those direct products for which the number of antisymmetric components is even. Conversely, for odd-length antisymmetric DFTs, the direct products with an odd number of antisymmetric components are chosen. The multiplicative complexity of each is then simply the sum of the complexities of each of the terms chosen for the union.

6.4. Discrete Hartley Transform

The discrete Hartley transform (DHT) proposed by Bracewell [7, 8, 9] is defined by

$$H_k = \sum_{n=0}^{N-1} x_n \text{cas}(2\pi nk/N), \quad k = 0, 1, \ldots, N-1,$$

where $\text{cas}(\theta) = \cos(\theta) + \sin(\theta)$. The major advantages of the DHT over the DFT are that a real-valued input sequence has a real-valued transform, and the forward and inverse transforms are identical except for a scaling by $1/N$ in the inverse transform.

The only difference between the defining formulas for the DFT and DHT is the "j" multiplying the sine term in the DFT. Therefore, we conclude that

$$H_k = \text{Re}[X_k] - \text{Im}[X_k]$$

and conversely that

$$\text{Re}[X_k] = \frac{1}{2}(H_{N-k} + H_k)$$

and

$$\text{Im}[X_k] = \frac{1}{2}(H_{N-k} - H_k),$$

where $H_N = H_0$ for notational convenience.

The existence of an invertible rational transformation between the DHT and DFT coefficients implies that the two systems are equivalent (in the sense of Definition 3.1), and therefore they have identical multiplicative complexities for the same length. This is expressed in the following corollary.

Corollary 6.9. $\mu_B(\text{DHT}(N); Q) = \mu_B(\text{DFT}(N); Q)$ *for all N*.

6.5. Discrete Cosine Transform The discrete cosine transform (DCT) of a sequence is defined by

$$c_k = \frac{2e_k}{N} \sum_{n=0}^{N-1} x_n \cos(\pi(2n+1)k/2N), \quad k = 0, 1, \ldots, N-1,$$

where

$$e_k = \begin{cases} 1/\sqrt{2} & \text{for } k = 0 \\ 1 & \text{otherwise.} \end{cases}$$

The inverse DCT (IDCT) is the transpose of the DCT operator, without the $2/N$ normalization, i.e.,

$$x_n = \sum_{k=0}^{N-1} e_k c_k \cos(\pi(2n+1)k/2N), \quad n = 0, 1, \ldots, N-1.$$

The normalizations by $2/N$ and $1/\sqrt{2}$ will be ignored in the further analysis since the $2/N$ scaling is rational and the $1/\sqrt{2}$ scaling is not important for most applications. For the remainder of this section all references to the DCT will be to the system obtained by omitting these scalings. The DCT has been found to be useful in data compression algorithms. It has been shown that the DCT has properties very similar to the Karhunen-Loève transform for first-order stationary Markov random data [2].

A common (albeit inefficient) technique for computing the DCT was due to the observation that a symmetric DFT also has only cosine terms, and that the DCT can be computed by forming the even extension of the data sequence, then using this extended sequence as the odd-indexed inputs to a quadruple-length FFT algorithm (usually the complex FFT available in some subroutine library). This is the opposite philosophy to that used in our symmetric DFT algorithms - we essentially discarded the redundant samples and computed only half of the outputs. This relation is useful, though, because it tells us that DCT(N) is really a part of DFT($4N$), as we could also have inferred from the arguments of the cosine functions.

From Chapter 5 we know that computation of the real and imaginary outputs of DFT($4N$) for any N are equivalent systems. Therefore we will examine the computation of the real output of DFT($4N$), which requires $\mu_B(\text{DFT}(4N); Q)/2$ m/d steps.

Thus

$$\text{Re}[X_k] = \sum_{n=0}^{4N-1} x_n \cos(\frac{2\pi nk}{4N}), \quad k = 0, 1, \ldots, 4N-1$$

$$= \sum_{n=0}^{2N-1} x_{2n} \cos(\frac{2\pi nk}{2N}) + \sum_{n=0}^{2N-1} x_{2n+1} \cos(\frac{\pi(2n+1)k}{2N})$$

$$= \sum_{n=0}^{2N-1} x_{2n} \cos(\frac{2\pi nk}{2N}) + \sum_{n=0}^{N-1} (x_{2n+1} + x_{N-2n-1}) \cos(\frac{\pi(2n+1)k}{2N}).$$

Let the two sums be denoted by $S_k = \sum_{n=0}^{2N-1} x_{2n} \cos(\frac{2\pi nk}{2N})$ and $T_k = \sum_{n=0}^{N-1} (x_{2n+1} + x_{N-2n-1}) \cos(\frac{\pi(2n+1)k}{2N})$. We need only compute S_k and T_k for the first N outputs, since $\text{Re}[X_k] = S_k + T_k$, $\text{Re}[X_{k+N}] = S_{N-k} - T_k$, $\text{Re}[X_{k+2N}] = \text{Re}[X_{2N-k}]$, and $\text{Re}[X_{k+3N}] = \text{Re}[X_{N-k}]$, $k = 0, 1, \ldots, N-1$. Computing S_k is exactly the computation of the real output of DFT($2N$) for the sequence x_{2n}, and T_k is DCT(N) for the sequence $x_{2n+1} + x_{N-2n-1}$.

This decomposition of the computation of $\text{Re}[\text{DFT}(4N)]$ into a union of disjoint systems satisfies the conditions of Theorem 3.9, yielding

$$\mu_B(\text{DFT}(4N; Q))/2 = \mu_B(\text{DFT}(2N); Q)/2 + \mu_B(\text{DCT}(N); Q).$$

Rearranging this gives the following expression for DCT(N).

Theorem 6.3. $\mu_B(\text{DCT}(N); Q) = \dfrac{\mu_B(\text{DFT}(4N); Q) - \mu_B(\text{DFT}(2N); Q)}{2}.$

From Theorem 6.3 we can deduce that for odd-length sequences

$$\mu_B(\text{DCT}(2K+1); Q) = \mu_B(\text{DFT}(2K+1); Q)$$

and for lengths that are powers of two,

$$\mu_B(\text{DCT}(2^r); Q) = 2^{r+1} - r - 2.$$

Formulas for other lengths can be easily derived using Theorems 5.4 and 6.3.

6.6. Summary of Chapter 6

In this chapter we have considered the multiplicative complexity of the DFT when only one output point is desired, when only one input point is nonzero, and when the input is symmetric or antisymmetric. The number of m/d steps necessary to compute one output of the DFT depends on the greatest common divisor of the output index and the sequence length. For the worst (and most common case) the complexity is on the order of $\phi(N)$, or slightly less than N. Computing more than one output can quickly approach the complexity of computing the entire DFT, depending on which output is selected.

Computation of the DFT from one sample is exactly the transpose of computing one output of the DFT, therefore the multiplicative complexity is identical.

When the input sequence is symmetric or antisymmetric, then the complexity is roughly half that of computing the full DFT. Computation of a symmetric DFT always requires fewer or the same number of m/d steps as the antisymmetric DFT.

APPENDIX A

Cyclotomic Polynomials and Their Properties

The cyclotomic polynomials are the monic polynomials with rational coefficients whose roots consist of a single occurrence of each of the primitive N^{th} roots of unity for some N. A cyclotomic polynomial can be expressed as a product of the (complex) linear polynomials contributed by each of its roots or as the quotient of the polynomial u^N-1, whose roots constitute all N^{th} roots of unity, with the product of the cyclotomic polynomials for all integers less than N that divide N. In the second representation, the product in the denominator contains all nonprimitive N^{th} roots of unity, thus the quotient has the effect of removing the nonprimitive roots from the set of all roots, leaving only the primitive roots. Let $C_N(u)$ be the Nth cyclotomic polynomial and w_N be a primitive N^{th} root of unity, then

$$C_N(u) = \prod_{\substack{0 \le i < N \\ (i,N)=1}} (u-w_N^i) = \frac{u^N-1}{\prod_{\substack{d \mid N \\ d \ne N}} C_d(u)}. \tag{A.1}$$

Another common notation for the Nth cyclotomic polynomial is $\phi_N(u)$, but this will not be used here to avoid confusion with the Euler function $\phi(N)$.

Several properties of cyclotomic polynomials are important in the study of the multiplicative complexity of the discrete Fourier transform. The first important property is that the cyclotomic polynomials are irreducible over Q, the field of rational numbers. This is a well-known property of cyclotomic polynomials and will not be proven here, a proof can be found in [34]. The degree of the Nth cyclotomic polynomial is equal to $\phi(N)$, the Euler function of N, defined to be the number of positive integers less than N and relatively prime to N. This is easily seen from the first representation of $C_N(u)$ in (A.1).

The coefficients of $C_N(u)$ are always integers, and it can be proven that the coefficients of $C_N(u)$ are elements of the set $\{-1, 0, 1\}$ when N has fewer than three distinct odd prime factors [24]. $C_N(u)$ is symmetric, except for $C_1(u)$. Some identities that are useful in generating the coefficients of particular cyclotomic polynomials are given in [24].

One characteristic of cyclotomic polynomials that is extremely important in the study of the multiplicative complexity of the DFT is how they factor in extension fields of the rationals, and in particular the rationals extended by roots of unity. This problem has been investigated by Winograd, at least to the point of determining the number of irreducible polynomial factors of $C_N(u)$ in extension fields of the rationals by roots of unity [45]. Winograd's result will be presented, and a version of his proof given, plus an example of the application of the result.

Theorem A.1. *The number of irreducible factors of the cyclotomic polynomial $C_b(u)$ in the field $Q(w_a)$ is $\phi((a,b))$, that is, the Euler function of the greatest common divisor of a and b.*

Proof. Let $w_n = e^{-j2\pi/n}$ be a primitive n^{th} root of unity, $d = (a,b)$ be the greatest common divisor of a and b, and $c = [a,b] = ab/d$ be the least common multiple of a and b. The following relations hold: $w_{ab}^d = w_c$, $w_{ab}^a = w_c^{a/d} = w_b$, and $w_{ab}^b = w_c^{b/d} = w_a$. Since the exponents a/d and b/d of w_c are integers, we have shown that $w_a, w_b \in Q(w_c)$ and therefore that $Q(w_a, w_b) \subseteq Q(w_c)$. Since $(a,b) = d$, we can express d as $d = ra + sb$ for some integers r and s. Using this relation we can show that $Q(w_c) = Q(w_a, w_b)$, since $w_c = w_{ab}^d = w_{ab}^{ra} w_{ab}^{sb} = (w_{ab}^a)^r (w_{ab}^b)^s = w_b^r w_a^s$ implies $w_c \in Q(w_a, w_b)$. For any two primitive n^{th} roots of unity, w_n and v_n, $Q(w_n) = Q(v_n)$. Let v_b be any primitive b^{th} root of unity, then $Q(w_c) = Q(w_a, w_b) = Q(w_b)(w_a) = Q(v_b)(w_a) = Q(w_a, v_b) = Q(w_a)(v_b)$, and the degree of any primitive b^{th} root of unity over $Q(w_a)$ is $[Q(w_c):Q(w_a)] = \phi(c)/\phi(a)$. Let P be any irreducible factor of $C_b(u)$ over $Q(w_a)$, then since the roots of P must be primitive b^{th} roots of unity, we conclude that $\deg P = \phi(c)/\phi(a)$. This implies that all irreducible polynomial factors of $C_b(u)$ over $Q(w_a)$ have degree $\phi(c)/\phi(a)$.

Let $F(a,b)$ be the number of irreducible factors of $C_b(u)$ in $Q(w_a)$, then $\phi(b) = F(a,b)\phi(c)/\phi(a)$ and $F(a,b) = \phi(a)\phi(b)/\phi(c)$. Note that a and b can be expressed in their canonical prime power factorizations as $a = \prod_i p_i^{e_{ia}}$ and $b = \prod_i p_i^{e_{ib}}$, where each of the products includes all prime factors from both a and b. From the definition, $c = \prod_i p_i^{\max(e_{ia}, e_{ib})}$ and $d = \prod_i p_i^{\min(e_{ia}, e_{ib})}$. The standard formula for determining $\phi(n)$ given the canonical factorization of n yields

$$\phi(a) = \prod_i p_i^{e_{ia}-1} \prod_{\substack{i \\ e_{ia} \neq 0}} (p_i - 1), \tag{A.2}$$

$$\phi(b) = \prod_i p_i^{e_{ib}-1} \prod_{\substack{i \\ e_{ib} \neq 0}} (p_i - 1), \tag{A.3}$$

$$\phi(c) = \prod_i p_i^{\max(e_{ia}, e_{ib})-1} \prod_i (p_i-1), \tag{A.4}$$

and

$$\phi(d) = \prod_i p_i^{\min(e_{ia}, e_{ib})-1} \prod_{\substack{i \\ \min(e_{ia}, e_{ib}) \neq 0}} (p_i-1). \tag{A.5}$$

It can be shown that $\phi(a)\phi(b) = \phi(c)\phi(d)$ by isolating each of the prime factors of a and b and noting that the contribution of p_i to $\phi(a)\phi(b)$ is $p_i^{e_{ia}+e_{ib}}(p_i-1)^2$ if $p_i|a$ and $p_i|b$ or it is $p_i^{e_{ia}+e_{ib}}(p_i-1)$ if only one of a and b is divisible by p_i, and the contribution of p_i to $\phi(c)\phi(d)$ is identical since the sum of the maximum and minimum of e_{ia} and e_{ib} must be equal to $e_{ia}+e_{ib}$ and whether one or two factors of p_i-1 are necessary again depends on whether just one or both of a and b are divisible by p_i. We conclude that $\phi(a)\phi(b)/\phi(c) = \phi(d)$ and thus that $F(a,b) = \phi(d) = \phi((a,b))$. ∎

As an example, Theorem A.1 will be applied to the irreducibility of cyclotomic polynomials over the rationals extended by j, the primitive fourth root of unity. For what positive integers N is $\phi((4,N)) \neq 1$? Since the only divisors of 4 are 1, 2, and 4, then $(4,N)$ can only have those values, and $\phi(1) = \phi(2) = 1$, $\phi(4) = 2$. Therefore, $C_N(u)$ is reducible in $Q(j)$ only when $(4,N) = 4$, that is, when $4|N$, and is irreducible over $Q(j)$ when $(4,N) \neq 4$. In this particular case it is easy to determine the irreducible factors of $C_N(u)$ in $Q(j)$.

Theorem A.2. *If* $N = 2^k m$, *where* m *is an odd integer, and* $k \geq 2$, *then* $C_N(u) = C_m(ju^{2^{k-2}})C_m(-ju^{2^{k-2}})$.

Proof. We will need one property of cyclotomic polynomials that is proved in [24]. The necessary property is that for any choice of integers m and p, $C_{mpk}(u) = C_{mp}(u^{p^{k-1}})$. Since m is odd, $(m,4) = 1$ and $[m,4] = 4m$, therefore the primitive $(4m)^{th}$ roots of unity are generated by the pairwise products of each of the primitive 4th roots of unity with each of the primitive m^{th} roots of unity. The Chinese remainder theorem guarantees that any non-negative integer less than $4m$ is uniquely representable as the sum $4r+ms \pmod{4m}$ where $0 \leq r < m$ and $0 \leq s < 4$, and if we restrict r and s such that $(r,m) = 1$, $(s,4) = 1$, these are the exponents of w_m and w_4, respectively, which are also primitive m^{th} and 4th roots of unity.

The product $w_m^r w_4^s$ is a primitive $(4m)^{th}$ root of unity, and since r takes on $\phi(m)$ values and s takes on $\phi(4)$ values, the total number of r and s pairs is $\phi(m)\phi(4) = \phi(4m)$, which is the total number of $(4m)^{th}$ roots of unity. These pairwise products generate the complete set of primitive $(4m)^{th}$ roots of unity since the resulting exponents of w_{4m} are distinct and the number of products is equal to the number of primitive $(4m)^{th}$ roots of unity. The primitive 4th roots of unity are j and $-j$, thus

122

if w_m is a root of $C_m(u)$, then jw_m must be a root of $C_m(-ju)$ and $-jw_m$ must be a root of $C_m(ju)$. The entire set of products of j and $-j$ with the roots of $C_m(u)$ was shown to comprise the set of primitive $(4m)^{th}$ roots of unity, therefore the roots of the product $C_m(ju)C_m(-ju)$ are the complete set of primitive $(4m)^{th}$ roots of unity, and since these same roots are, by definition, roots of the $(4m)^{th}$ cyclotomic polynomial, then $C_{4m}(u) = C_m(ju)C_m(-ju)$. If u is replaced by u to some power, then the identity still holds, and therefore $C_{2^k m}(u) = C_{4m}(u^{2^{k-2}}) = C_m(-ju^{2^{k-2}})C_m(ju^{2^{k-2}})$. ∎

APPENDIX B

Complexities of Multidimensional Cyclic Convolutions

Table B.1 tabulates the number of m/d steps that are necessary and sufficient to compute multidimensional cyclic convolutions for all possible combinations of lengths requiring 100 or fewer m/d steps. The table is first ordered by the number of m/d steps, then by the total number of points being convolved. The remaining columns are the lengths in each dimension.

Table B.1. Number of m/d steps (μ) necessary
to compute multidimensional cyclic convolutions.

μ	$\prod N_i$	N_1	N_2	N_3	N_4	N_5	N_6
4	4	2	2				
8	6	2	3				
8	8	2	2	2			
10	8	2	4				
13	9	3	3				
16	10	2	5				
16	12	2	2	3			
16	12	2	6				
16	16	2	2	2	2		
18	12	3	4				
20	16	2	2	4			
22	16	4	4				
24	14	2	7				
24	16	2	8				
26	15	3	5				
26	18	2	3	3			
26	18	3	6				
30	18	2	9				
32	20	2	2	5			
32	20	2	10				
32	24	2	2	2	3		

μ	$\prod N_i$	N_1	N_2	N_3	N_4	N_5	N_6
32	24	2	2	6			
32	32	2	2	2	2	2	
34	20	4	5				
36	24	2	3	4			
36	24	2	12				
36	24	4	6				
38	21	3	7				
40	22	2	11				
40	24	3	8				
40	27	3	3	3			
40	32	2	2	2	4		
43	25	5	5				
44	32	2	4	4			
46	27	3	9				
48	26	2	13				
48	28	2	2	7			
48	28	2	14				
48	32	2	2	8			
50	28	4	7				
50	32	4	8				
52	30	2	3	5			
52	30	2	15				
52	30	3	10				
52	30	5	6				
52	36	2	2	3	3		
52	36	2	3	6			
52	36	6	6				
54	32	2	16				
57	36	3	3	4			
57	36	3	12				
60	36	2	2	9			
60	36	2	18				
62	33	3	11				
63	36	4	9				
64	34	2	17				
64	40	2	2	2	5		
64	40	2	2	10			
64	48	2	2	2	2	3	
64	48	2	2	2	6		
64	64	2	2	2	2	2	2
66	35	5	7				

μ	$\prod N_i$	N_1	N_2	N_3	N_4	N_5	N_6
68	40	2	4	5			
68	40	2	20				
68	40	4	10				
72	38	2	19				
72	40	5	8				
72	48	2	2	3	4		
72	48	2	2	12			
72	48	2	4	6			
74	39	3	13				
76	42	2	3	7			
76	42	2	21				
76	42	3	14				
76	42	6	7				
76	48	3	4	4			
76	48	4	12				
80	44	2	2	11			
80	44	2	22				
80	45	3	3	5			
80	45	3	15				
80	48	2	3	8			
80	48	2	24				
80	48	6	8				
80	54	2	3	3	3		
80	54	3	3	6			
80	64	2	2	2	2	4	
82	44	4	11				
84	45	5	9				
86	48	3	16				
86	50	2	5	5			
86	50	5	10				
88	46	2	23				
88	64	2	2	4	4		
89	49	7	7				
92	54	2	3	9			
92	54	3	18				
92	54	6	9				
92	64	4	4	4			
94	50	2	25				
96	52	2	2	13			
96	52	2	26				
96	56	2	2	2	7		

μ	$\prod N_i$	N_1	N_2	N_3	N_4	N_5	N_6
96	56	2	2	14			
96	64	2	2	2	8		
98	51	3	17				
98	52	4	13				
100	54	2	27				
100	56	2	4	7			
100	56	2	28				
100	56	4	14				
100	64	2	4	8			

APPENDIX C

Programs for Computing Multiplicative Complexity

The following set of Fortran routines evaluates the functions necessary to compute the multiplicative complexity of the one-dimensional DFT. These include an interactive main program that prompts for the DFT length and returns the computed complexity, a subroutine to compute the complexity of multidimensional cyclic convolutions, subroutines for factoring integers and computing all divisors of an integer, and functions for evaluating Euler's totient function and the greatest common divisor of two integers. The decomposition of the DFT into a set of multidimensional cyclic convolutions as described in Theorem 5.4 is used for determining the complexity.

The program will work for lengths N larger than this limit but N must then have no prime factors greater than 1531^2 and at most one non-repeated prime factor between 1531 and 1531^2. These limitations may be removed by changing the way that the factoring subroutine works or increasing the maximum prime in its internal table. Other limitations are that the number of distinct prime factors of N must be less than ten, the maximum number of divisors of N is 1000, and the maximum number of divisors of $\phi(d)$ for any $d \mid N$ is also 1000. The smallest integer with ten distinct prime factors is 6,469,693,230 which is slightly larger than 2,147,483,647, the largest positive integer representable on 32-bit computers using two's complement arithmetic. The smallest integer with more than 1000 divisors is 245,044,800 which has 1008 divisors.

```
C    DFTMUL - MULTIPLICATIVE COMPLEXITY OF 1-D DFT
C
C    DFTMUL INTERACTIVELY REQUESTS THE SEQUENCE LENGTH
C    FOR WHICH THE MULTIPLICATIVE COMPLEXITY OF THE DFT
C    IS TO BE COMPUTED, COMPUTES THE COMPLEXITY, AND THEN
C    OUTPUTS THE NUMBER OF REQUIRED MULTIPLICATIONS.
C
C    REQUIRED SUBPROGRAMS ARE:
C         MULTCC -  MULTIDIMENSIONAL CYCLIC
C                   CONVOLUTION COMPLEXITY
C         FACTOR -  FACTOR AN INTEGER
C         DIVISR -  DETERMINE POSITIVE DIVISORS
C         IGCD   -  GREATEST COMMON DIVISOR
C         IPHI   -  EULER'S (TOTIENT) FUNCTION
C
      INTEGER IP(10), IE(10), LISTD(1000)
      INTEGER IPDIV(10), IEDIV(10), NI(100)
100   CONTINUE
      WRITE (6, *)
      WRITE (6, *) 'N (0 TO STOP) ?'
      READ (5, *) N
      IF (N .EQ. 0) GO TO 400
      CALL FACTOR (N, IP, IE, NF)
      CALL DIVISR (N, IP, IE, NF, LISTD, NDIV)
      NMUL = 0
      DO 300 I = 1, NDIV
           ID4 = IGCD(LISTD(I), 4)
           IF (ID4 .EQ. 4) THEN
                NB4PHI = LISTD(I)/2
                NUMCC = 2
           ELSE
                NB4PHI = LISTD(I)
                NUMCC = 1
           ENDIF
           IF (NB4PHI .NE. 1) THEN
                CALL FACTOR (NB4PHI, IPDIV, IEDIV, NFDIV)
                DO 200 J = 1, NFDIV
                     NI(J) = (IPDIV(J)-1)*IPDIV(J)**(IEDIV(J)-1)
200             CONTINUE
           ELSE
                NI(1) = 1
                NFDIV = 1
           ENDIF
```

```
            NMUL = NMUL + NUMCC*(MULTCC(NI,NFDIV) - 1)
300  CONTINUE
     WRITE (6, *) 'NUMBER OF MULTIPLICATIONS IS ',NMUL
     GO TO 100
400  STOP
     END
```

```
C
C     MULTCC - NUMBER OF MULTIPLICATIONS NECESSARY FOR
C              MULTIDIMENSIONAL CYCLIC CONVOLUTION.
C
C     INPUTS ARE NDIM, THE NUMBER OF DIMENSIONS, AND AN
C     ARRAY N CONTAINING THE LENGTH OF THE CYCLIC
C     CONVOLUTION IN EACH OF THE DIMENSIONS.
C     A RETURNED VALUE OF -1 INDICATES BAD INPUT PARAMETERS.
C
C     REQUIRED SUBPROGRAMS ARE:
C          FACTOR - FACTOR AN INTEGER
C          DIVISR - DETERMINE POSITIVE DIVISORS
C          IGCD   - GREATEST COMMON DIVISOR
C          IPHI   - EULER'S (TOTIENT) FUNCTION
C
      INTEGER FUNCTION MULTCC(N,NDIM)
      INTEGER IP(10,10), IE(10,10)
      INTEGER LISTD(100,10), IPHIDV(100,10)
      INTEGER N(10), NF(10), NDIV(10)
      INTEGER IPDUM(10), IEDUM(10)
      INTEGER ICOUNT(10), INUM(10), IDEN(10), LCM(10)
      MULTCC = -1
      IF (NDIM .LT. 0) RETURN
      MULTCC = 2
      DO 300 I = 1, NDIM
           IF (N(I) .LT. 0) GO TO 100
           MULTCC = MULTCC * N(I)
           CALL FACTOR (N(I), IP(1,I), IE(1,I), NF(I))
           CALL DIVISR (N(I), IP(1,I), IE(1,I), NF(I),
     1            LISTD(1,I), NDIV(I))
           DO 200 J = 1, NDIV(I)
                CALL FACTOR (LISTD(J,I), IPDUM,
     1               IEDUM, NFDUM)
                IPHIDV(J,I) = IPHI(LISTD(J,I),
     1               IPDUM, IEDUM, NFDUM)
200        CONTINUE
300   CONTINUE
      NUMTOT = 1
      LCMTOT = 1
      DO 400 I = 1, NDIM-1
           ICOUNT(I) = 1
           INUM(I) = IPHIDV(1,I)
           NUMTOT = NUMTOT * INUM(I)
```

```
              IDEN(I) = LISTD(1,I)
              LCM(I) = LCMTOT * IDEN(I)/IGCD(LCMTOT,IDEN(I))
              LCMTOT = LCM(I)
  400  CONTINUE
       INDEX = NDIM
  500  CONTINUE
       IF (INDEX .EQ. NDIM) THEN
              DO 600 I = 1, NDIV(NDIM)
                     LCMD = LCMTOT * LISTD(I,NDIM)/
     1                     IGCD(LCMTOT,LISTD(I,NDIM))
                     CALL FACTOR (LCMD, IPDUM, IEDUM, NFDUM)
                     IDNTOT = IPHI(LCMD, IPDUM, IEDUM, NFDUM)
                     MULTCC = MULTCC - IPHIDV(I,NDIM) *
     1                     NUMTOT/IDNTOT
  600         CONTINUE
              INDEX = INDEX - 1
       ELSE
              IF (ICOUNT(INDEX) .NE. 0)
     1               NUMTOT = NUMTOT/INUM(INDEX)
              ICOUNT(INDEX) = ICOUNT(INDEX) + 1
              IF (ICOUNT(INDEX) .GT. NDIV(INDEX)) THEN
                     ICOUNT(INDEX) = 0
                     INDEX = INDEX - 1
              ELSE
                     INUM(INDEX) = IPHIDV(ICOUNT(INDEX),INDEX)
                     NUMTOT = NUMTOT*INUM(INDEX)
                     IDEN(INDEX) = LISTD(ICOUNT(INDEX),INDEX)
                     IF (INDEX .GT. 1) THEN
                            LCM(INDEX) = LCM(INDEX-1)*IDEN(INDEX)/
     1                            IGCD(LCM(INDEX-1),IDEN(INDEX))
                     ELSE
                            LCM(INDEX) = IDEN(INDEX)
                     ENDIF
                     LCMTOT = LCM(INDEX)
                     INDEX = INDEX + 1
              ENDIF
       ENDIF
       IF (INDEX .NE. 0) GO TO 500
  700  RETURN
  100  CONTINUE
       MULTCC = -1
       RETURN
       END
```

```
C
C     FACTOR - DETERMINE THE PRIME POWER DECOMPOSITION OF N.
C
C     NF IS THE NUMBER OF DISTINCT PRIME FACTORS, IP IS AN
C     ARRAY CONTAINING THE PRIME FACTORS OF N, AND IE IS
C     AN ARRAY CONTAINING THE EXPONENT OF THE CORRESPONDING
C     ENTRY IN IP.
C
      SUBROUTINE FACTOR (N,IP1,IE1,NF)
      PARAMETER (NP = 243)
      DIMENSION IP(NP), IP1(1), IE1(1)
      DATA IP/2,3,5,7,11,13,17,19,23,29,31,37,41,43,47,53,59,61,67,71,
     . 73,79,83,89,97,101,103,107,109,113,127,131,137,139,149,151,157,
     . 163,167,173,179,181,191,193,197,199,211,223,227,229,233,239,241,
     . 251,257,263,269,271,277,281,283,293,307,311,313,317,331,337,347,
     . 349,353,359,367,373,379,383,389,397,401,409,419,421,431,433,439,
     . 443,449,457,461,463,467,479,487,491,499,503,509,521,523,541,547,
     . 557,563,569,571,577,587,593,599,601,607,613,617,619,631,641,643,
     . 647,653,659,661,673,677,683,691,701,709,709,719,727,733,739,743,
     . 751,757,761,769,773,787,797,809,811,821,823,827,829,839,853,857,
     . 859,863,877,881,883,887,907,911,919,929,937,941,947,953,967,971,
     . 977,983,991,997,1009,1013,1019,1021,1031,1033,1039,1049,1051,
     . 1061,1063,1069,1087,1091,1093,1097,1103,1109,1117,1123,1129,
     . 1151,1153,1163,1171,1181,1187,1193,1201,1213,1217,1223,1229,
     . 1231,1237,1249,1259,1277,1279,1283,1289,1291,1297,1301,1303,
     . 1307,1319,1321,1327,1361,1367,1373,1381,1399,1409,1423,1427,
     . 1429,1433,1439,1447,1451,1453,1459,1471,1481,1483,1487,1489,
     . 1493,1499,1511,1523,1531/
      N1 = N
      NF = 0
      IF (N .EQ. 1) RETURN
      DO 100 I = 1, NP
          MD = N1/IP(I)
          IF (MD*IP(I) .NE. N1) GO TO 100
          N1 = MD
          NF = NF + 1
          IP1(NF) = IP(I)
          IE1(NF) = 1
200       CONTINUE
          MD = N1/IP(I)
          IF (MD*IP(I) .NE. N1) GO TO 300
          N1 = MD
          IE1(NF) = IE1(NF)+1
```

```
                GO TO 200
300             CONTINUE
                IF (N1 .EQ. 1) RETURN
100  CONTINUE
     IF (N1 .LT. IP(NP)*IP(NP)) THEN
                NF = NF + 1
                IP1(NF) = N1
                IE1(NF) = 1
                RETURN
     ELSE
                WRITE (6,*) N,' HAS A FACTOR GREATER THAN ',IP(NP)
                STOP
     ENDIF
     END
```

```
C
C     DIVISR - DETERMINE ALL THE POSITIVE DIVISORS OF N.
C
C     INPUTS ARE THE NUMBER N, ITS DISTINCT PRIME FACTORS
C     IN IP, THEIR EXPONENTS IN IE, AND THE NUMBER OF
C     DISTINCT PRIME FACTORS NF.  OUTPUTS ARE NDIV, THE
C     NUMBER OF DIVISORS OF N AND LISTD, A LIST OF THE
C     DIVISORS OF N.
C
      SUBROUTINE DIVISR (N,IP,IE,NF,LISTD,NDIV)
      DIMENSION IP(1), IE(1), LISTD(1)
      NDIV = 1
      LISTD(NDIV) = 1
      DO 300 I = 1, NF
            NDOLD = NDIV
            MUL = 1
            DO 200 J = 1, IE(I)
                  MUL = MUL * IP(I)
                  DO 100 K = 1, NDOLD
                        NDIV = NDIV + 1
                        LISTD(NDIV) = LISTD(K) * MUL
100               CONTINUE
200         CONTINUE
300   CONTINUE
      RETURN
      END
```

```
C
C     IGCD - GREATEST COMMON DIVISOR OF TWO INTEGERS
C
C     IGCD USES THE EUCLIDEAN ALGORITHM TO OBTAIN THE
C     GREATEST COMMON DIVISOR, (I1,I2), OF THE TWO INPUT
C     INTEGERS, I1 AND I2, AND RETURNS THE RESULT IN IGCD.
C
C     MIKE HEIDEMAN
C     JULY 26, 1984
C
      INTEGER FUNCTION IGCD(I1,I2)
      N1 = MAX(ABS(I1),ABS(I2))
      N2 = MIN(ABS(I1),ABS(I2))
      IF (N2 .EQ. 0) GO TO 300
100   CONTINUE
      NR = MOD(N1,N2)
      IF (NR .EQ. 0) GO TO 200
      N1 = N2
      N2 = NR
      GO TO 100
200   CONTINUE
      IGCD = N2
      RETURN
300   CONTINUE
      IGCD = N1
      RETURN
      END
```

```
C
C     IPHI - COMPUTE EULER'S PHI FUNCTION AT N
C
C     INPUTS ARE THE NUMBER N, ITS DISTINCT PRIME FACTORS
C     IN IP, THEIR EXPONENTS IN IE, AND THE NUMBER OF
C     DISTINCT PRIME FACTORS NF.
C
      INTEGER FUNCTION IPHI(N,IP,IE,NF)
      DIMENSION IP(1), IE(1)
      IPHI = 1
      DO 100 I = 1, NF
          IPHI = IPHI * (IP(I)-1) * (IP(I) ** (IE(I)-1))
100   CONTINUE
      RETURN
      END
```

APPENDIX D

Tabulated Complexities of the One-Dimensional DFT

Table D.1 shows the number of m/d steps that are necessary and sufficient to compute the one-dimensional discrete Fourier transform over the field of rational numbers for the first 1000 positive integer lengths. The input is assumed to be real-valued. For complex inputs the number of m/d steps is twice the value indicated in the table.

N	μ	N	μ	N	μ	N	μ	N	μ
1	0	41	71	81	136	121	226	161	293
2	0	42	46	82	142	122	214	162	272
3	1	43	75	83	159	123	215	163	313
4	0	44	60	84	92	124	204	164	284
5	4	45	56	85	139	125	227	165	262
6	2	46	78	86	150	126	162	166	318
7	7	47	87	87	149	127	239	167	327
8	2	48	38	88	124	128	198	168	192
9	8	49	82	89	167	129	227	169	316
10	8	50	74	90	112	130	192	170	278
11	15	51	80	91	137	131	251	171	279
12	4	52	68	92	156	132	188	172	300
13	17	53	97	93	155	133	219	173	337
14	14	54	74	94	174	134	246	174	298
15	14	55	86	95	158	135	210	175	294
16	10	56	60	96	106	136	212	176	266
17	26	57	89	97	179	137	263	177	335
18	16	58	98	98	164	138	238	178	334
19	29	59	111	99	157	139	267	179	351
20	16	60	56	100	148	140	184	180	224
21	23	61	107	101	190	141	263	181	341
22	30	62	102	102	160	142	262	182	274
23	39	63	81	103	195	143	247	183	323
24	12	64	84	104	140	144	166	184	316
25	37	65	96	105	142	145	256	185	323
26	34	66	94	106	194	146	262	186	310
27	37	67	123	107	207	147	249	187	340
28	28	68	104	108	148	148	248	188	348
29	49	69	119	109	203	149	289	189	293
30	28	70	92	110	172	150	228	190	316
31	51	71	131	111	188	151	287	191	371
32	32	72	70	112	138	152	236	192	264
33	47	73	131	113	213	153	258	193	369
34	52	74	124	114	178	154	250	194	358
35	46	75	114	115	206	155	270	195	292
36	32	76	116	116	196	156	212	196	328
37	62	77	125	117	173	157	299	197	382
38	58	78	106	118	222	158	294	198	314
39	53	79	147	119	204	159	293	199	383
40	36	80	86	120	120	160	206	200	302

Table D.1. Number of m/d steps (μ) necessary to compute a length-N discrete Fourier transform for real-valued inputs.

N	μ	N	μ	N	μ	N	μ	N	μ
201	371	241	459	281	543	321	623	361	700
202	380	242	452	282	526	322	586	362	682
203	367	243	449	283	555	323	600	363	681
204	320	244	428	284	524	324	544	364	548
205	366	245	432	285	478	325	568	365	668
206	390	246	430	286	494	326	626	366	646
207	373	247	429	287	525	327	611	367	723
208	298	248	412	288	392	328	572	368	650
209	379	249	479	289	559	329	629	369	669
210	284	250	454	290	512	330	524	370	646
211	403	251	491	291	539	331	643	371	703
212	388	252	324	292	524	332	636	372	620
213	395	253	477	293	577	333	578	373	731
214	414	254	478	294	498	334	654	374	680
215	390	255	421	295	566	335	630	375	685
216	304	256	438	296	500	336	430	376	700
217	377	257	502	297	521	337	651	377	697
218	406	258	454	298	578	338	632	378	586
219	395	259	452	299	559	339	641	379	739
220	344	260	384	300	456	340	556	380	632
221	390	261	467	301	545	341	609	381	719
222	376	262	502	302	574	342	558	382	742
223	435	263	519	303	572	343	657	383	759
224	318	264	384	304	494	344	604	384	608
225	371	265	496	305	550	345	622	385	662
226	426	266	438	306	516	346	674	386	738
227	447	267	503	307	599	347	687	387	697
228	356	268	492	308	500	348	596	388	716
229	443	269	529	309	587	349	683	389	769
230	412	270	420	310	540	350	588	390	584
231	379	271	523	311	611	351	585	391	748
232	396	272	438	312	432	352	574	392	662
233	455	273	415	313	607	353	691	393	755
234	346	274	526	314	598	354	670	394	764
235	446	275	476	315	462	355	670	395	750
236	444	276	476	316	588	356	668	396	628
237	443	277	539	317	625	357	616	397	773
238	408	278	534	318	586	358	702	398	766
239	467	279	481	319	599	359	711	399	661
240	274	280	376	320	476	360	460	400	630

Table D.1. (Continued)

N	μ	N	μ	N	μ	N	μ	N	μ
401	784	441	770	481	862	521	1023	561	1024
402	742	442	780	482	918	522	934	562	1086
403	727	443	875	483	883	523	1031	563	1119
404	760	444	752	484	904	524	1004	564	1052
405	716	445	846	485	906	525	888	565	1076
406	734	446	870	486	898	526	1038	566	1110
407	760	447	869	487	959	527	996	567	995
408	648	448	712	488	860	528	814	568	1052
409	799	449	881	489	941	529	1042	569	1127
410	732	450	742	490	864	530	992	570	956
411	791	451	829	491	967	531	1021	571	1123
412	780	452	852	492	860	532	876	572	988
413	797	453	863	493	934	533	991	573	1115
414	746	454	894	494	858	534	1006	574	1050
415	806	455	726	495	838	535	1046	575	1094
416	646	456	720	496	850	536	988	576	892
417	803	457	895	497	945	537	1055	577	1130
418	758	458	886	498	958	538	1058	578	1118
419	827	459	832	499	987	539	1007	579	1109
420	568	460	824	500	908	540	840	580	1024
421	815	461	907	501	983	541	1055	581	1133
422	806	462	758	502	982	542	1046	582	1078
423	805	463	907	503	999	543	1025	583	1127
424	780	464	810	504	660	544	910	584	1052
425	775	465	814	505	963	545	1030	585	922
426	790	466	910	506	954	546	830	586	1154
427	773	467	927	507	951	547	1075	587	1167
428	828	468	692	508	956	548	1052	588	996
429	745	469	881	509	1009	549	989	589	1095
430	780	470	892	510	842	550	952	590	1132
431	851	471	899	511	937	551	1049	591	1148
432	650	472	892	512	932	552	960	592	1022
433	843	473	897	513	887	553	1049	593	1173
434	754	474	886	514	1004	554	1078	594	1042
435	772	475	874	515	990	555	973	595	1047
436	812	476	816	516	908	556	1068	596	1156
437	835	477	899	517	1005	557	1105	597	1151
438	790	478	934	518	904	558	962	598	1118
439	867	479	951	519	1013	559	1039	599	1187
440	696	480	638	520	776	560	798	600	924

Table D.1. (Continued)

N	μ	N	μ	N	μ	N	μ	N	μ
601	1175	641	1263	681	1343	721	1385	761	1503
602	1090	642	1246	682	1218	722	1400	762	1438
603	1129	643	1275	683	1355	723	1379	763	1439
604	1148	644	1172	684	1116	724	1364	764	1484
605	1152	645	1174	685	1326	725	1368	765	1329
606	1144	646	1200	686	1314	726	1362	766	1518
607	1203	647	1283	687	1331	727	1439	767	1495
608	1042	648	1098	688	1234	728	1104	768	1330
609	1105	649	1269	689	1321	729	1408	769	1517
610	1100	650	1136	690	1244	730	1336	770	1324
611	1183	651	1135	691	1363	731	1404	771	1508
612	1032	652	1252	692	1348	732	1292	772	1476
613	1205	653	1297	693	1179	733	1451	773	1537
614	1198	654	1222	694	1374	734	1446	774	1394
615	1102	655	1270	695	1350	735	1302	775	1418
616	1008	656	1162	696	1200	736	1342	776	1436
617	1215	657	1201	697	1316	737	1425	777	1360
618	1174	658	1258	698	1366	738	1338	778	1538
619	1227	659	1307	699	1367	739	1463	779	1491
620	1080	660	1048	700	1176	740	1292	780	1168
621	1169	661	1295	701	1381	741	1291	781	1489
622	1222	662	1286	702	1170	742	1406	782	1496
623	1197	663	1174	703	1284	743	1475	783	1467
624	910	664	1276	704	1224	744	1248	784	1362
625	1214	665	1150	705	1342	745	1456	785	1510
626	1214	666	1156	706	1382	746	1462	786	1510
627	1141	667	1295	707	1360	747	1453	787	1563
628	1196	668	1308	708	1340	748	1360	788	1528
629	1185	669	1307	709	1403	749	1469	789	1559
630	924	670	1260	710	1340	750	1370	790	1500
631	1235	671	1237	711	1345	751	1483	791	1523
632	1180	672	974	712	1340	752	1418	792	1268
633	1211	673	1319	713	1377	753	1475	793	1463
634	1250	674	1302	714	1232	754	1394	794	1546
635	1214	675	1208	715	1272	755	1454	795	1492
636	1172	676	1264	716	1404	756	1172	796	1532
637	1159	677	1342	717	1403	757	1487	797	1585
638	1198	678	1282	718	1422	758	1478	798	1322
639	1209	679	1277	719	1431	759	1435	799	1564
640	1056	680	1120	720	986	760	1272	800	1332

Table D.1. (Continued)

N	μ	N	μ	N	μ	N	μ	N	μ
801	1533	841	1660	881	1739	921	1799	961	1894
802	1568	842	1630	882	1540	922	1814	962	1724
803	1537	843	1631	883	1745	923	1779	963	1885
804	1484	844	1612	884	1560	924	1516	964	1836
805	1502	845	1602	885	1702	925	1735	965	1856
806	1454	846	1610	886	1750	926	1814	966	1766
807	1589	847	1623	887	1767	927	1777	967	1915
808	1524	848	1578	888	1512	928	1670	968	1814
809	1607	849	1667	889	1693	929	1843	969	1804
810	1432	850	1550	890	1692	930	1628	970	1812
811	1599	851	1648	891	1667	931	1741	971	1931
812	1468	852	1580	892	1740	932	1820	972	1796
813	1571	853	1691	893	1747	933	1835	973	1889
814	1520	854	1546	894	1738	934	1854	974	1918
815	1582	855	1470	895	1766	935	1727	975	1710
816	1330	856	1660	896	1544	936	1396	976	1746
817	1551	857	1703	897	1681	937	1847	977	1941
818	1598	858	1490	898	1762	938	1762	978	1882
819	1281	859	1699	899	1731	939	1823	979	1909
820	1464	860	1560	900	1484	940	1784	980	1728
821	1627	861	1579	901	1750	941	1867	981	1847
822	1582	862	1702	902	1658	942	1798	982	1934
823	1635	863	1719	903	1639	943	1837	983	1959
824	1564	864	1402	904	1708	944	1802	984	1728
825	1434	865	1696	905	1724	945	1566	985	1923
826	1594	866	1686	906	1726	946	1794	986	1868
827	1643	867	1680	907	1803	947	1883	987	1891
828	1492	868	1508	908	1788	948	1772	988	1716
829	1637	869	1689	909	1742	949	1763	989	1929
830	1612	870	1544	910	1452	950	1748	990	1676
831	1619	871	1663	911	1803	951	1877	991	1955
832	1392	872	1628	912	1498	952	1640	992	1766
833	1582	873	1637	913	1797	953	1887	993	1931
834	1606	874	1670	914	1790	954	1798	994	1890
835	1646	875	1647	915	1654	955	1870	995	1934
836	1516	876	1580	916	1772	956	1868	996	1916
837	1525	877	1739	917	1785	957	1801	997	1979
838	1654	878	1734	918	1664	958	1902	998	1974
839	1671	879	1733	919	1819	959	1869	999	1800
840	1152	880	1438	920	1656	960	1452	1000	1824

Table D.1. (Continued)

Problems

Chapter 2

2.1 The following algorithm computes the system $x_0y_0+x_1y_1$: $h_1 = x_0y_1$, $h_2 = y_1 \cdot y_1$, $h_3 = h_1 \cdot y_0$, $h_4 = x_1 \cdot h_2$, $h_5 = h_3 + h_4$, $h_6 = h_5/y_1$. Convert this to a division-free algorithm using Theorem 2.8.

2.2 In the proof of Theorem 2.8 it was stated that denominators had to have non-zero constant terms in order for the power series to exist. Show why this is a necessary condition.

2.3 The algorithm $m_0 = x_0y_0$, $m_1 = x_1y_1$, $m_2 = (x_0+x_1)(y_0+y_1)$, $z_0 = m_1$, $z_1 = m_2-m_1-m_3$, $z_2 = m_3$ is a minimal algorithm for computing the coefficients of $z(u) = (x_0+x_1u)(y_0+y_1u) = z_0+z_1u+z_2u^2$. A transpose of this system computes $z_0' = x_0'y_0'+x_1'y_1'$ and $z_1' = x_0'y_1'+x_1'y_2'$. Transpose the given algorithm to obtain an algorithm that computes the transposed system using 3 multiplications.

2.4 Propose a set of two specializations in the original system of Problem 2.3 that reduces the number of required m/d steps to zero.

2.5 Propose a set of three specializations in the transposed system of Problem 2.3 that reduces the number of required m/d steps to zero.

2.6 Convert the algorithm A of Example 2.3 into a noncommutative algorithm using the procedure of Theorem 2.9. Derive a minimal noncommutative algorithm for this system.

2.7 Prove the first half of the direct sum conjecture for the system $z_0 = x_0y_0$, $z_1 = x_1y_1$.

2.8 An algorithm is known for multiplying a pair of arbitrary 2×2 matrices using only 7 multiplications. If this algorithm is successively iterated using the tensor product formulation, how many multiplications are used in computing the product of $2^n \times 2^n$ matrices? How many multiplications are used in the straightforward method?

2.9 In Problem 2.8, the 7 multiplication algorithm requires 18 additions. How many additions are used by the tensor product algorithm for $2^n \times 2^n$ matrix products? How many additions are used in the straightforward method?

2.10 Propose a technique for multiplication of pairs of $N \times N$ matrices (N arbitrary) based on the 7 multiplication 2×2 product. Derive estimates for the number of

143

multiplications and additions required by this algorithm.

Chapter 3

3.1 Show that none of the entries of the matrix U' encountered in the proof of Theorem 3.2 can be zero when $k = 1$.

3.2 Derive the second algorithm of Example 3.2 using first the transpose approach and then Winograd's second method.

3.3 Derive an algorithm for complex multiplication using Winograd's third method based on the isomorphism $\sigma: G[u]/\langle u^2+1 \rangle \rightarrow G[v]/\langle v^2-2v+1 \rangle$ where $\sigma(a+bu) = (a-b)+bv$ and $\sigma^{-1}(c+dv) = (c+d)+du$. Could the resulting algorithm have been derived by Winograd's first method alone? Why or why not?

3.4 Derive a minimal algorithm for the direct product of systems presented in Example 2.5. Is this algorithm more efficient than just using the direct product of minimal algorithms for the subsystems?

3.5 Derive a complete algorithm that computes the system of in 7 multiplications.

3.6 Demonstrate a tensor product algorithm for computing the system $MPM(u^4+u^2+1)$ over Q using 9 multiplications and 3 constants from Q. Compare the total operations of this algorithm to that of a minimal algorithm.

3.7 Let $z(u) = z_0+z_1u+z_2u^2+z_3u^3$. Demonstrate the Chinese remainder theorem reduction and reconstruction of this polynomial over the irreducible factors (over Q) of u^4-1.

3.8 Derive an algorithm for computing the system of Example 3.1 using the constants $0, 1, -1, 2$. Convert this to an algorithm that uses the constants $\infty, 1, -1, \frac{1}{2}$ by considering the equivalent system $(x_1+x_0u)(y_2+y_1u+y_2u^2)$.

3.9 Demonstrate a factorization of $C_{36}(u)$ over $Q(w_{45})[u]$, where w_{45} is a primitive 45th root of unity. Conversely, demonstrate a factorization of $C_{45}(u)$ over $Q(w_{36})[u]$.

3.10 Determine the multiplicative complexity of a polynomial product modulo u^3-2, v^3-2 over Q. Reevaluate the complexity over $Q(2^{1/3})$.

Chapter 4

4.1 Show that multiplying a polynomial over Q by the quadratic $\sqrt{2}u^2+(\sqrt{6}+\sqrt{2})u+2+\sqrt{6}$ is equivalent to a product with a linear polynomial.

4.2 Show why the inequality $\sum_{j=1}^{r} m_j < m - (n+1)(r-1)$ must be satisfied for

$\mu_B(xy; G) < m+n+1$ when $x(u) = Q(u) + \sum\limits_{j=0}^{r} \sum\limits_{i=0}^{m_j} x_{ij}' u^i Q_i(u)$ where $Q(u), Q_i(u) \in G[u]$,

$i = 1, 2, \ldots, r$.

4.3 Prove that 1 is a root of all antisymmetric polynomials.

4.4 Prove that dividing an antisymmetric polynomial $x(u)$ by $u-1$ yields a symmetric polynomial.

4.5 Derive a minimal algorithm for computing the product $(x_0 + x_1 u + x_1 u^2 + x_0 u^3)(y_0 + y_1 u)$ over Q.

4.6 Derive a minimal algorithm for computing the product $(x_0 + x_1 u + x_1 u^2 + x_0 u^3)(y_0 + y_1 u + y_1 u^2 + y_0 u^3)$ over Q.

4.7 Derive a minimal algorithm over Q for computing the decimated polynomial product $PM(2, 2, 2)$ whose outputs are $z_0 = x_0 y_0$, $z_2 = x_0 y_2 + x_1 y_1 + x_2 y_0$, and $z_4 = x_2 y_2$.

4.8 Derive a minimal algorithm over Q for computing the coefficients z_0, z_1, and z_2 where $\sum\limits_{k=0}^{4} z_i u^i = (\sum\limits_{i=0}^{3} x_i u^i)(\sum\limits_{j=0}^{1} y_i u^i)$.

4.9 Derive an expression for $\mu_B(z_0, z_1, \ldots, z_r; G)$ when $x(u)$ is a symmetric polynomial.

4.10 Derive an expression for $\mu_B(z_0, z_1, \ldots, z_r; G)$ when $x(u)$ is an antisymmetric polynomial.

Chapter 5

5.1 Derive a minimal algorithm for DFT(3) over Q.

5.2 Show the decomposition of DFT(9) into a union of cyclic convolutions of lengths 1, 2, and 6. (Include the rational part of the system.)

5.3 Derive a minimal algorithm for DFT(9) over Q.

5.4 Derive a minimal algorithm for DFT(16) over Q.

5.5 For a primitive root g in the reduced residue system modulo p^i, $i \geq 2$, prove that $g^{p^{i-1} - p^{i-2}} = 1 + p^{i-1} n_i$, where $p \nmid n_i$.

5.6 Show that $\sin \dfrac{2\pi 5^n}{2^i} = \cos \dfrac{2\pi 5^{n+2^{i-4}}}{2^i}$ holds for $i \geq 5$ and all n.

5.7 Derive $\mu_B(DFT(3^b 5^c); Q)$.

5.8 Determine the least integer N for which at least one of the denominators $\phi([d_1, d_2, \ldots, d_m])$ in Theorem 5.4 is greater than one.

5.9   Derive $\mu_B(\text{DFT}(2^a 7^d); Q)$.

5.10   In the last step of the derivation of Example 5.1 show that the terms in the delta function sum are indeed zero for $i = k$, $k+1$, and $k+2$.

Chapter 6

6.1   Derive a minimal algorithm over Q for computing the real and imaginary components of X_7, the eighth output (counting from zero), of DFT(15).

6.2   Derive a formula for $\mu_B(\text{DCT}(N); Q)$ by substituting expressions for DFT complexities from Theorem 5.4 into Theorem 6.3.

6.3   Compute $\mu_B(\text{DFT}_s(315); Q)$ and $\mu_B(\text{DFT}_a(315); Q)$.

6.4   Derive an algorithm for DFT(7) that uses an identical structure for the computation of the symmetric and antisymmetric parts.

6.5   Prove that for DFT lengths such that $(N, 4) = 2$, $\mu_B(\text{DFT}_s(N); Q) = 2\mu_B(\text{DFT}_s(N/2); Q)$ and $\mu_B(\text{DFT}_a(N); Q) = 2\mu_B(\text{DFT}_a(N/2); Q)$.

6.6   Computation of DFT(N; 0) requires no m/d steps. Characterize all N and k for which $\mu_B(\text{DFT}(N; k); Q) = 0$.

6.7   For $N > 2$ and real inputs, show that $\mu_B(\text{DFT}(N; 1); Q)$ is an even number when N is divisible by 4 and odd otherwise.

6.8   Derive an expression for $\mu_B(\text{Re}[\text{DFT}(N; k)]; Q)$, i.e., the computation of the real part of the kth output of a length-N DFT for real-valued inputs.

6.9   Derive an expression for $\mu_B(\text{Im}[\text{DFT}(N; k)]; Q)$, i.e., the computation of the imaginary part of the kth output of a length-N DFT for real-valued inputs.

6.10   What is the minimum number of rational multiplications necessary to compute DFT(N; k) in a minimal algorithm over Q. If multiplications by integers greater than or equal to two are to be implemented as rational multiplications, rather than additions, then how many rational multiplications are necessary?

Bibliography

[1]   Agarwal, R.C. and Cooley, J.W., "New algorithms for digital convolution,"
       IEEE Trans. Acoust., Speech, Signal Processing, Vol. ASSP-25, no. 5,
       pp. 392-410, Oct. 1977.

[2]   Ahmed, N., Natarajan, T., and Rao, K.R., "Discrete cosine transform," *IEEE
       Trans. Comput.*, Vol. C-23, no. 1, pp. 90-93, Jan. 1974.

[3]   Auslander, L., Feig, E., and Winograd, S., "New algorithms for the multidi-
       mensional discrete Fourier transform," *IEEE Trans. Acoust., Speech, Signal
       Processing*, Vol. ASSP-31, no. 2, pp. 388-403, Apr. 1983.

[4]   Auslander, L., Feig, E., and Winograd, S., "Abelian semi-simple algebras and
       algorithms for the discrete Fourier transform," *Adv. in Appl. Math.*, Vol. 5,
       no. 1, pp. 31-55, Mar. 1984.

[5]   Auslander, L., Feig, E., and Winograd, S., "The multiplicative complexity of
       the discrete Fourier transform," *Adv. in Appl. Math.*, Vol. 5, no. 1, pp. 87-109,
       Mar. 1984.

[6]   Auslander, L. and Winograd, S., "The multiplicative complexity of certain
       semilinear systems defined by polynomials," *Adv. in Appl. Math.*, Vol. 1,
       no. 3, pp. 257-299, 1980.

[7]   Bracewell, R.N., "Discrete Hartley transform," *J. Opt. Soc. Amer.*, Vol. 73,
       no. 12, pp. 1832-1835, Dec. 1983.

[8]   Bracewell, R.N., "The fast Hartley transform," *Proc. IEEE*, Vol. 72, no. 8,
       pp. 1010-1018, Aug. 1984.

[9]   Bracewell, R.N., *The Hartley Transform.* New York: Oxford University
       Press, 1986.

[10]   Brockett, R.W. and Dobkin, D., "On the optimal evaluation of a set of bilinear
       forms," *Linear Algebra Appl.*, Vol. 19, no. 3, pp. 207-235, 1978.

[11]   Burrus, C.S. and Eschenbacher, P.W., "An in-place, in-order prime factor FFT
       algorithm," *IEEE Trans. Acoust., Speech, Signal Processing*, Vol. ASSP-29,
       no. 4, pp. 806-817, Aug. 1981.

147

[12]   Burrus, C.S. and Parks, T.W., *DFT/FFT and Convolution Algorithms*. New York: John Wiley and Sons, 1985.

[13]   Chu, S. and Burrus, C. S., "A Prime Factor FFT Algorithm Using Distributed Arithmetic," *IEEE Trans. Acoust., Speech, Signal Processing*, Vol. ASSP-30, no. 2, pp. 217-227, Apr. 1982.

[14]   Cook, S.A., "On the minimum computation time of functions," PhD Thesis, Harvard Univ., Cambridge, MA, 1966.

[15]   Cooley, J.W. and Tukey, J.W., "An algorithm for the machine calculation of complex Fourier series," *Math. Comput.*, Vol. 19, no. 90, pp. 297-301, Apr. 1965.

[16]   Feig, E., "Minimal algorithms for bilinear forms may have divisions," *J. Algorithms*, Vol. 4, no. 1, pp. 81-84, Mar. 1983.

[17]   Fiduccia, C.M. and Zalcstein, Y., "Algebras having linear multiplicative complexities," *J. ACM*, Vol. 24, no. 2, pp. 311-331, Apr. 1977.

[18]   Goertzel, G., "An algorithm for the evaluation of finite trigonometric series," *Am. Math. Monthly*, Vol. 65, no. 1, pp. 34-35, Jan. 1958.

[19]   Goldstine, H.H., *A History of Numerical Analysis from the 16th Through the 19th Century*. New York, Heidelberg, and Berlin: Springer-Verlag, 1977.

[20]   Good, I.J., "The interaction algorithm and practical Fourier analysis," *J.R. Statist. Soc. B*, Vol. 20, no. 2, pp. 361-372, 1958., Addendum in *J.R. Statist. Soc. B*, Vol. 22, no. 2, pp. 372-375, 1960..

[21]   Heideman, M.T., Johnson, D.H., and Burrus, C.S., "Gauss and the history of the fast Fourier transform," *Archive for History of Exact Sciences*, Vol. 34, no. 3, pp. 265-277, 1985.

[22]   Karatsuba, A. and Ofman, Yu., "Multiplication of multidigit numbers on automata," *Soviet Physics Dokl.*, Vol. 7, pp. 595-596, 1963.

[23]   Kolba, D.P. and Parks, T.W., "A prime factor FFT algorithm using high-speed convolution," *IEEE Trans. Acoust., Speech, Signal Processing*, Vol. ASSP-25, no. 4, pp. 281-294, Aug. 1977.

[24]   McClellan, J.H. and Rader, C.M., *Number Theory in Digital Signal Processing*. Englewood Cliffs, NJ: Prentice-Hall, 1979.

[25]   Motzkin, T.S., "Evaluation of polynomials and evaluation of rational functions," *Bull. Amer. Math. Soc.*, Vol. 61, p. 163, 1955.

[26]   Niven, I. and Zuckerman, H.S., *An Introduction to the Theory of Numbers*. New York: Wiley, 1980.

[27]   Nussbaumer, H.J., *Fast Fourier Transform and Convolution Algorithms*. Berlin, Heidelberg, and New York: Springer-Verlag, 1981.

[28]   Oppenheim, A. V. and Schafer, R. W., *Digital Signal Processing*. Englewood Cliffs, NJ: Prentice-Hall, 1975.

[29]   Ostrowski, A.M., *On Two Problems in Abstract Algebra Connected with Horner's Rule, Studies Presented to R. von Mises*. New York: Academic Press, 1954.

[30]   Pan, V.Ya., "Some schemes for the calculation of polynomials with real coefficients," *Dokl. Akad. Nauk SSSR*, Vol. 127, no. 2, pp. 266-269, 1959.

[31]   Pan, V.Ya., "Certain schemes for the evaluation of polynomials with real coefficients," *Problems of Cybernetics*, Vol. 5, pp. 14-32, 1961.

[32]   Rabin, M.O. and Winograd, S., "Fast evaluation of polynomials by rational preparation," *Comm. Pure App. Math.*, Vol. 25, no. 4, pp. 433-458, Jul. 1972.

[33]   Rader, C.M., "Discrete Fourier transforms when the number of data samples is prime," *Proc. IEEE*, Vol. 56, no. 6, pp. 1107-1108, Jun. 1968.

[34]   Sah, C.-H., *Abstract Algebra*. New York: Academic Press, 1967.

[35]   Silverman, H.F., "An introduction to programming the Winograd Fourier transform algorithm (WFTA)," *IEEE Trans. Acoust., Speech, Signal Processing*, Vol. ASSP-25, no. 2, pp. 152-165, Apr. 1977.

[36]   Strassen, V., "Gaussian elimination is not optimal," *Numer. Math.*, Vol. 13, pp. 354-356, 1969.

[37]   Strassen, V., "Vermeidung von divisionen," *J. Reine Angew. Math.*, Vol. 264, pp. 184-202, 1973.

[38]   Thomas, L.H., "Using computers to solve problems in physics," in *Applications of Digital Computers*, W.F. Freiberger and W. Prager, Eds. Ginn: Boston, MA, pp. 44-45, 1963.

[39]   Toom, A.L., "The complexity of a scheme of functional elements realizing the multiplication of integers," *Soviet Math. Dokl.*, Vol. 4, no. 3, pp. 714-716, May-Jun. 1963.

[40]   Winograd, S., "On the number of multiplications necessary to compute certain functions," *Comm. Pure App. Math.*, Vol. 23, no. 2, pp. 165-179, Mar. 1970.

[41]   Winograd, S., "The effect of the field of constants on the number of multiplications," *Proc. 16th Ann. Symp. Foundations of Computer Science*, pp. 1-2, Berkeley, CA, Oct. 13-15, 1975.

[42]   Winograd, S., "On computing the discrete Fourier transform," *Proc. Nat. Acad. Sci. USA*, Vol. 73, no. 4, pp. 1005-1006, Apr. 1976.

[43]   Winograd, S., "Some bilinear forms whose multiplicative complexity depends on the field of constants," *Math. Syst. Theory*, Vol. 10, no. 2, pp. 169-180, 1977.

[44]   Winograd, S., "On computing the discrete Fourier transform," *Math. Comput.*, Vol. 32, no. 141, pp. 175-199, Jan. 1978.

[45]   Winograd, S., "On the multiplicative complexity of the discrete Fourier transform," *Advances in Math.*, Vol. 32, no. 2, pp. 83-117, May 1979.

[46]   Winograd, S., "On multiplication in algebraic extension fields," *Theoret. Comput. Sci.*, Vol. 8, no. 3, pp. 359-377, Jun. 1979.

[47]   Winograd, S., *Arithmetic Complexity of Computations*.   Philadelphia, PA: SIAM Publications, 1980.

[48]   Winograd, S., "On multiplication of polynomials modulo a polynomial," *SIAM J. Comput.*, Vol. 9, no. 2, pp. 225-229, May 1980.

[49]   Yip, P.C.Y., "Some aspects of the zoom transform," *IEEE Trans. Comput.*, Vol. C-25, no. 3, pp. 287-296, Mar. 1976.

Index